D0955376

Because I *Fly*

BECAUSE I

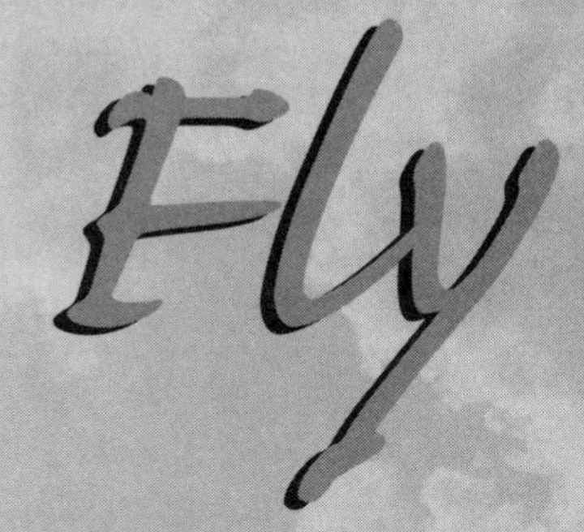

A Collection of Aviation Poetry

Selected and Compiled by
HELMUT REDA

McGRAW-HILL

New York • Chicago • San Francisco • Lisbon • London
Madrid • Mexico City • Milan • New Delhi
San Juan • Seoul • Singapore • Sydney • Toronto

McGraw-Hill

A Division of The ***McGraw-Hill*** *Companies*

1 2 3 4 5 6 7 8 9 0 DOC/DOC 0 7 6 5 4 3 2 1

ISBN 0-07-138085-X

The sponsoring editor for this book was Shelley Ingram Carr, the editing supervisor was Daina Penikas, and the production supervisor was Pamela A. Pelton. It was designed and set in Baskerville and Caflish Script by Jeff Potter of Potter Publishing Studio in Shelburne Falls, Massachusetts.

Printed and bound by R. R. Donnelley & Sons Company.

This book was printed on recycled, acid-free paper containing a minimum of 50% recycled, de-inked fiber.

Dedication

TO THE MEN, WOMEN, AND FAMILIES of the United States Armed Services who proudly serve our country. Their sacrifices helped make us a great nation and advanced democracy throughout the world.

To Helmut Sr., Elfriede, George, Angela, and Jeneen: Thanks for your unconditional love and support while I pursued my flying interests. Time spent aloft was time spent away from you. These poems will help you understand my relentless quest for flight and why I made certain sacrifices.

To Alex, Loren, Crystal, Nichole, and Jessica: Aim high, the future is yours for the making. Find your own niche of happiness and pursue it vigorously! Almost everything is possible, so choose wisely!

To Ted Wierzbanowski (W+12), Robert Barthelemy, Vance Brand, Tom Hedgecock, Nathan Jones, Dale Brown, and Jimmy Doolittle III: Your leadership, character, and values inspired my own.

Contents

The Classics

Flight: Solitude, Freedom, Beauty, Mystery, Meaning, and Motivation

Pilots

Training and Solo

Grounded

Religion/Prayer

Military Service

Soaring

For Children

The Physical Sky

Those Who Make It Happen

Historical Feats of Wrights, Lindbergh, and Earhart

Pushing the Envelope

FLIGHT TEST

ASTRONAUTS AND SPACE

Air Racing

Death by Flying

Stand By

Acknowledgments

THIS ANTHOLOGY was developed via cyberspace by a team who've never met face-to-face and eventually became friends. Our common bond was a deep appreciation for aviation poetry and without their help this anthology would have never been published. I would like to provide my most sincere thank you to: Shelley Carr, Daina Penikas, Michael Larkin, Major Jeff Alfier, Hy Yarchun, Dr. Barbara Hendrickson, and Kathleen Rodgers. I would also like to thank Leanne Sparr; Kathy Feltovich; Julie Wullkotte; Joanne Rumple; Dr. Walt McDonald; Dr. John Clark Pratt; Peter Jensen; Dick Sims; Jack Nichols; Dave English; Frank and Karen Deaton; Christine Neff; James Ellis; Tony Burton; Dr. Don Anderson; Allan and Barbara Stanley; Eric and Karl Stice; Jimmy Driskell; Keith Rosenkranz; Dale Brown; Brigadier General Robin Olds, USAF (Ret); Ana Arteaga; Erwin Doreleijer; Royal Netherlands Air Force Lt Kols Ton Vosters, Herman "BLEEP" Koolstra and Harry Horlings; Barbara Bell; Mal "RAMBO" Tutty; Major Bill Hack; Major Marlon Camacho; Captain Patricia Rodriquez-Rey; Phil Harris; Lt Col Bob Ireland; Larry "WILDMAN" Shope; Betty Rice; and Daniel de Zendegui for their friendship; humor, encouragement, kind words, editorial, and research assistance.

Finally, I gratefully thank all poets and publishers for permission to reprint their copyrighted material. Every reasonable effort was made to find the copyright holder of all copyrighted poems in this book. Any errors are inadvertent and will be corrected or attributed in future editions provided notification is sent to the publisher.

Listed alphabetically by poet or source:

Major Jeffrey C. Alfier for "Anamnesis Of A Bombardier: My Father, Sleepwandering From The War," "Gaining Altitude," and "Precision Guided Munitions – Fourteen Second Flight To Dispossession," first time published in this Anthology.

A. P. Watt Ltd on behalf of Michael B. Yeats for "An Irish Airman Foresees his Death" by W. B. Yeats.

Doug Atkins for "To All Aircrew," first time published in this anthology.

Richard Bach for quote used in Foreword, *A Gift of Wings,* p. 237

Richard L. Barlow for "The Seat" and "The First Time," first time published in this anthology.

Timothy S. Bastian for "A Dance Through the Sky," "Why I Fly," and "The Days That We Have Flown," first time published in this anthology.

Carolyn Berge for "Feeling Compassionate," first time published in this anthology.

Desmond M. Chorley for "To Fly", reprinted from *Thoughts Take Flight: An Anthology of Poetry and Stories About Airplanes, Pilots, and Flying,* compiled and edited by Allan and Barbara Stanley, and published by CAVU Press, 1986.

The Ciardi Family Publishing Trust, John L. Ciardi, Trustee, for "P-51," first time printed in the June 1944 issue of *Poetry*, "First Snow on an Airfield" first printed in the June 1944 issue of *Poetry,* "Return" first printed in his 1988 book *Saipan,* "The Pilot in the Jungle" first published in his 1949 book *Live Another Day,* and "Visibility Zero" first printed in his 1988 book *Saipan.*

Patricia and Michael Collins for "To A Husband Who Must Seek The Stars," reprinted from *Carrying the Fire,* published by Farrar, Straus and Giroux in 1974.

Dirk Elber and courtesy of *Soaring Magazine* for "Soaring Verse," reprinted from *Soaring Magazine* in August 1988.

James A. Grimshaw for "To Have Been for Tony Dater 14 January 1974."

Frederick T. Kiley for Untitled.

Dr. John Clark Pratt for "From A Pilot: For Walton F. Dater," reprinted from the1974 issue of United States Air Force Academy *Icarus* magazine.

James A. Grimshaw for "March 10, 1966," first time published in this anthology.

Ivan L. Fail for "Tribute To A Queen," "When The Mustangs Came," "The Liberator," and "Today I Rode A Legend."

Free Flight, Canada's Soaring Magazine, for the permission to reprint "Rotor Winds" and "Lesson Six is Done," by Tom Schollie.

Dr. James Freeman for "Goal," "Nick," "The Glider Pilot's Lament," and "The Path We Choose," published by the *Oz Report* in January 2000.

The United States Air Force *Fighter Weapons Review* for "A Distant Thunder," by Dallas Blevins, reprinted in the Spring 1985 issue.

Flying Safety magazine for "The Troop Who Rode One In" by unknown author reprinted in the December 1984 issue.

Clay Greager for "The Aviator Man."

Audrey V. Greene for "Soaring Soliloquy," "A Place To Play," "Dreams," "Parallels," "Gems Of Wisdom," "It's Easy To Forget," "My Thermal," "Absurd Possibilities," "Will I Ever Know," "Myriad Scenes of Soaring," and "The Sad Sack," by Jack P. Greene, reprinted from *Sunshine and Shadows of Soaring* published by the Aero Club Albatross in 1984 and not copyrighted.

Graciela C. Griener for "To a Pilot's Wife," by her late husband Gene Griener, first published in his book, *The Knife is Wood,* copyright 1978 by Jewel Books Diamond Publishing Company.

Patrick Hamilton for "Cloud Dreamers," first time published in this anthology.

Hansel Herrera for "The Line Mechanic," first time published in this anthology.

Daniel Hibbard and courtesy of *Soaring Magazine* for "The Cloud," first published in September 1992 *Soaring Magazine.*

Judy Humphrey for "Last Flight," first time published in this anthology.

United States Air Force Academy *Icarus* magazine and Dr. John Clark Pratt for "From a Pilot: For Walton F. Dater" by Lieutenant Colonel John Clark Pratt, reprinted from the 1974 issue.

Mary von Schrader Jarrell and Boston University for Randall Jarrell's "The Death of the Ball Turret Gunner," printed in the Winter 1945 issue of *Partisan Review.*

Dr. Scott A. Jenkins and courtesy of *Soaring Magazine* for "Downburst," reprinted from the July 1987 issue of *Soaring Magazine*.

H. Gene Johnson for "The More True Love," first time published in this anthology.

Richard Kerti for "To a Pilot's Son," reprinted from the April 1995 issue of *Air Line Pilot Magazine*.

Fredrick T. Kiley for "A Platform and aa Passion" by Ronald E. Pedro, "How Close Formation" by Don Clelland, "Words for Don Morris" by John Clark Pratt, "And I Look Down" by Roland Herwig, "When It's Over" by James H. Tiller III, "The Wingmen From Takhli" by Eugene Cirillo, "Untitled" by Whitney I. Blair, "Death Of A Flyer" by David J. Smith, "To A Pilot's Widow" by Eileen Lundin, and "Dog Fight" by John J. Kowalski, reprinted from *Listen, The War: A Collection of Poetry about the Viet-Nam War,* Edited by Lieutenant Colonel Fred Kiley and Lieutenant Colonel Tony Dater.

Preston F. Kirk for "The Leap," reprinted from *Thoughts Take Flight: An Anthology of Poetry and Stories about Airplanes, Pilots, and Flying,* compiled and edited by Allan and Barbara Stanley, and published by CAVU Press, 1986.

Michael J. Larkin for "Flying West" (first published by the *Airline Pilot Magazine* February 1995), "Reno Races" (first published by TARPA Magazine); "The Aviator," and "War Stories;" none of the poems were copyrighted.

Peter H. Liotta and Cleveland State University Poetry Center for "U-2," first time published in *Onionhead,* "To The Man I Never Met," first time published in *Rainy Day,* and "Dead Reckoning," first time published in *Risely Review,* all three poems copyrighted by Peter H. Liotta in 1991.

James MacNutt for "Gift," "On Top," and "Westbound Seven Four," all first time published in this anthology, and "Sunrise Flight," which was published by the American Poetry Society in 1988.

Elizabeth MacKethan Magid for "Celestial Flight."

Daniel C. McCorry, Jr. for "Mountain," reprinted from the January 1975 issue of *Talon* magazine.

Walt McDonald's poems "Night Solo, Georgia," "The Flying Dutchman," "For Dawes, On Takeoff," "Praying A Stall Won't Spin Us," and "Buzzing in a Biplane," reprinted from *The Flying Dutchman,* Ohio State University Press, 1987; reprinted by permission of the author.

Walt McDonald and University of Notre Dame Press for "Old Pilots in the Crowd at Kitty Hawk," from *All Occasions,* copyright 2000 by University of Notre Dame Press, used by permission of the publisher and author.

P. A. Monahan for "The Quiet," reprinted from *Thoughts Take Flight: An Anthology of Poetry and Stories About Airplanes, Pilots, and Flying,* compiled and edited by Allan and Barbara Stanley, and published by CAVU Press, 1986.

P. A. Monahan for "Seventh Heaven," copyright 1993.

David A. Morris for "Come and Fly," "Wings to Fly," and "Ode to My Hero," first time published in this anthology.

Peggy Nemerov for "The War in the Air," first published in the *Paris Review,* 1986.

New Statesman & Society and *The Observer* for "To Poets and Airmen," by Stephen Spender.

Dave Nichols for "No Stalling," first time published in this anthology.

Gary Osoba for "Van Gogh Sky," "Escape!" and "The Surpassing Way of the Sky," first time published in this anthology.

David Pedlow for "Evening Flight," "Viewed From Another Angle," and "Winter

High," first time published in this Anthology.

John Clark Pratt for "For Don Morris: a T-28 pilot KIA, May, 1970"; "From a Pilot: for Walton F. Dater, Lt Col USAF (1932–1974)"; and "Vapor Trails–1."

Jeannette Nordquist Purviance for "Her First Solo," reprinted from *Thoughts Take Flight: An Anthology of Poetry and Stories About Airplanes, Pilots, and Flying,* compiled and edited by Allan and Barbara Stanley, and published by CAVU Press, 1986.

C. W. "Bill" Getz for "Phantom" by Major Paul, "Ode To A Bombardier," by unknown author, "The Fighter Pilot – A Tribute," by Raymond B. Tucker, and "The Return," by unknown author, reprinted from *The Wild Blue Yonder: Songs of the Air Force, Volume II, Stag Bar Edition* published by The Redwood Press in 1986 and "An Airman's Hymn" by Francis M. Miller, reprinted in *Wild Blue Yonder: Songs of the Air Force,* published by The Redwood Press in 1981.

Dave Ray for "For the Guys on the Ground" and "Just Another Flying Day," first time published in this anthology.

Patricia Rockwell for "Air Race," reprinted from *Thoughts Take Flight, An Anthology of Poetry and Stories About Airplanes, Pilots, and Flying,* compiled and edited by Allan and Barbara Stanley, and published by CAVU Press, 1986; "Those Who Fly," and "Small Planes And Dark Clouds" first time published in this anthology.

Kathleen M. Rodgers for "Because You Have Flown," To Live to Fly," "The Searcher," "Last of the Breed…," "An Old Head," and "A Little Boy's Dream," all first time published in this anthology, and "The Lady Let Him Fly," reprinted from the Spring 1992 issue of *Daedalus Flyer.*

Victoria Schrauwen for "Free," first time published in this anthology.

Betty Simpson for "First Solo Flight," first time published in *Tacoma News Tribune Sunday Magazine* in February 1979.

Ford H. Smart for "What Is A Fighter Pilot?" reprinted in *Wild Blue Yonder: Songs of the Air Force,* published by The Redwood Press in 1981.

The Society of Experimental Test Pilots for "Test Pilot," by Gill Robb Wilson.

Misty Sorensen for "An Airman's Blessing" and "A Nation Cried," first time published in this anthology.

Eric Stice for "Thermal," first time published in this anthology.

Karl Stice for "Sky Fever," first time published in this anthology.

Geoffrey H. Tyler for "The Choice," copyright 1982.

Patty Wagstaff for "To All Pilots," first published in her book *Fire and Ice: A Life on the Edge,* by Chicago Review Press in 1997

Wesleyan University Press for permission to reprint "Two Poems of Flight-Sleep," "The Liberator Explodes," and "The Firebombing" by James Dickey from his book *The Whole Motion: Collected Poems 1945–1992.*

Gill Robb Wilson's poems were reprinted from *Leaves From An Old Log,* by Gill Robb Wilson, published by American Aviation Associates, Washington, D.C., 1938.

Wright-Patterson AFB Education Fund for "What is a Test Pilot" and "Wright Field, 1944: The Ready Room," both by unknown authors, reprinted from *Test Flying at Old Wright Field,* stories collected by Ken Chilstrom and edited by Penn Leary, copyright 1991 by Wright Patterson AFB Educational Fund.

Chris Woods for "The Voice," first time published in this anthology.

Sources of poems by unknown or anonymous poets not referenced above: "Silly Old Baboon" was found on the Canadian Aviation Poetry Web site (http://www.midwintercanada.com), developed by Stewart Midwinter, an active, internationally competitive hang glider pilot. He came upon this poem while riding on a train and talking to Terence Alan Milligan (born April 16, 1918). Milligan published the poem in his 1968 book, A *Book of Milliganimals.* "The Eagles and Black Nights" was written by someone from the 57th Fighter Interceptor

Squadron, Keflavik, Iceland. The poem was found on a plaque in the Base Exchange at Wright-Patterson AFB, Ohio. "The Forgotten Man," "Wings of My Silver Plane," "Tilly," "840 by Accident," "Ode to a P-51," "A Fighter Pilot's Friend" were found in the historical archives poetry folder M6-a at the United States Air Force Museum, Wright-Patterson AFB, OH. "Because I Fly" was found on a plaque hung on the wall at the 3247th Test Squadron Stag Bar, Eglin AFB, FL. "Why I Want to Be a Pilot" was created by a fifth grader and first published in *South Carolina Aviation News*. The Air Force Hymn was found on the government Web site http://www.mer.cap.gov/hymn.htm. The Air Force Psalm was found in the historical archives poetry folder at the United States Air Force Museum, Wright-Patterson AFB, OH with the Office of Chief of Air Force Chaplain's stationary and seal affixed to it. "The Parable of Joe" was printed in the *Alberta Soaring Council ASCent,* March 1995 issue and found by Tony Burton, editor of Canadian Soaring Magazine, Free Flight in a 30-year-old Royal Canadian Air Force magazine possibly imported from the Royal Air Force.

About the Cover

COVER PHOTO is the Sikorsky S-40 designed by Igor Ivanovitch Sikorsky during 1930 and served as the first Pan American Clipper from 1931 until 1944. The S-40 was America's first flying boat heralding the era of large ocean-spanning flying boats in the early 1930s and remains a treasured part of the American legacy. *Historical research by Major Hi Yarchun, United States Air Force Reserve.*

Preface

THIS BOOK DOES NOT BEGIN with this sentence. It started over 30 years ago, when I first became fascinated with flight and starting building model rockets and gliders. By the age of 14, I mastered the hobby and set a United States record in model rocketry. I continued with my passion by proudly joining the United States Air Force in 1980. After obtaining my single engine land, sea, and glider ratings I started reading aviation poetry to reminisce my experiences.

Through my research I discovered there were no current, all encompassing, sources of aviation poetry readily available. Previous aviation poetry books were hard to obtain, outdated, and sometimes difficult to understand. I researched, collected, and studied over 115 books and magazines from the United States, the United Kingdom, Canada, France, and Australia. From this effort I developed a comprehensive bibliography of aviation poetry books. I include this bibliography to document and archive the evolution of aviation poetry.

The purpose of this book is to consolidate the "best of the best" in aviation poetry and make it readily available to others. I reviewed over 500 aviation poems using a subjective selection criteria to determine which poems belonged in this collection. The poems had to 1) be easily read and experienced; 2) flow smoothly through rhyme and meter; 3) transmit significant meaning and emotion; 4) possess excellent imagery, symbolism, and tone; and 5) surpass other poetry in total impact. I was primarily interested in the emotional response and message rather than the structural or contextual format. To comprehend the relative value of these poems, I dedicated the first section of the Anthology to the classic poets: James Dickey, W.B. Yeats, Randall Jarrell, John Ciardi, Howard Nemerov, and Sir Stephen Spender. I also used the most popular poem "High Flight" by

John Gillespie Magee as the first poem in the Anthology to be the normative standard to assess the other poems in the book.

The collection contains 176 poems from 75 poets covering the period between 1869 and 2001. It is the most complete, up-to-date, collection of aviation poetry ever published! The book is carefully subdivided into 15 topic areas so readers can quickly access poems of specific interest. For instance, there are 14 poems from renowned classic poets, 16 poems about pilots, 15 poems about flight, 17 poems about death, and 8 poems set-aside just for children. Professional and recreational pilots wrote over 90 percent of the poems. Besides the classic poets, there were other notable poets including Dr. Walt McDonald, Texas State Poet Laureate for 2001; Patty Wagstaff, three-time U.S. National Aerobatics Champion; and Gary Osoba, a glider World Record holder. Based on my research, Walt McDonald published more aviation poems than any other poet totaling 120 poems. Gill Robb Wilson (1893 to 1966) published over 50 aviation poems and was probably one of the best known and most well-rounded aviation writers of our time.

I compiled these poems for the many souls who are deeply moved by flight. This collection of poetry serves as an exciting introduction to the experience of flight for those armchair pilots who have never flown. These poems will fascinate your children to a whole new world and inspire them to learn more about flight. For the pilot, keeping this book in your flight bag offers a safe alternative to pressing ahead in stormy IFR conditions, when good judgment dictates not leaving the ground. These poems will take your spirit aloft while you sit comfortably inside watching a crackling fire. This collection will also help others understand "who you are" and why you do things in a different way. Through these pages, your spouse, children, or loved ones will see how flight molds a certain type of personal discipline that allows survival in a very unforgiving environment. They will understand why you are so calculating, and how you balance between your conservative and daring nature. These poems will help others understand your need to live on the edge, to explore the unknown, and to do what few have done before. If you are truly intrigued by flight, this book is for you!

My description of this collection is complete, the foundation set, and the concepts constructed. It is your book now! Read, reflect, and reminisce about your own flying experiences. If I have brought back pleasant memories or added peace and tranquility to your life, my efforts have not been wasted!

Introduction

"Rare is the man who has been exposed to the intense heat of a pilot's enthusiasm, without in some way being affected by it. The only reason that this can be is the unreasonable itself, that strange distant mystique of machines that carry men through the air."

–Richard Bach

"Once you have flown, you will walk the earth with your eyes turned skyward, for there you have been, and there you long to return."

–Leonardo da Vinci

THERE IS SOMETHING ABOUT the mystique of flight that stirs the soul and dazzles the imagination. From the beginning of mankind's existence, we have longed to fly. Our prehistoric ancestors dreamed of mighty wings to gain vantage in their evolutionary struggles. Inspired by flight's unencumbered freedom, they endeavored to liberate themselves from two-dimensional confinement. After centuries of ponder, tinker, toil and trial, mankind conquered flight's mystery. What was freely given to nature, mankind stole through technological application. Powerful, man-made, rigid machines replaced nature's anatomical perfection. Through brute force and creative ingenuity we transcended our terrestrial presence to venture the skies. Mankind's destiny has forever changed and now lies with the stars.

Now that we fly, we can escape from our hectic domestic gridiron infrastructure to the peaceful sanctuary of the sky. Urban congestion, degeneration, and pollution slowly disappear as we climb to altitude. From up on high, the earth is viewed as whole and humanity's presence is blurred across nature's scenic beauty. It is from this perspective, God's perspective, we begin to understand the immensity of our domain.

Flight exposes us to new sensations almost beyond the realm of expression. We experience the entire continuum between the serene tranquility felt while soaring, to the death defying exhilaration and terror of air-to-air combat. Many of us have lost close personal friends due to the dangers of flight. It is through these unique perspectives, sensations, and experience that aviation poetry finds its roots and song. Poetry serves as a bridge to help us eloquently express to others what flight is all about.

I've been interested in flight since childhood. When flying became too expensive, I read aviation poetry to maintain continuity and to satisfy my yearnings to fly. Although poetry may not replicate our flying experiences, it helps us reminisce and appreciate how fortunate we are to have flown. This collection of poems captures mankind's spirit and helps us understand the beauty, lore and mystique of flight.

Helmut Reda

United States Mission to the United Nations
Geneva, Switzerland
July 2001

BECAUSE I *Fly*

The Classics

High Flight

Oh, I have slipped the surly bonds of earth
And danced the skies on laughter-silvered wings;
Sunward I've climbed, and joined the tumbling mirth
Of sun-split clouds–and done a hundred things
You have not dreamed of–wheeled and soared and swung
High in sunlit silence. Hov'ring there,
I've chased the shouting wind along, and flung
My eager craft through footless halls of air.
Up, up the long, delirious, burning blue
I've topped the windswept heights with easy grace
Where never lark, or even eagle flew.
And, while with silent, lifting mind I've trod
The high untrespassed sanctity of space,
Put out my hand, and touched the face of God

–John Gillespie Magee

The Death of the Ball Turret Gunner

From my mother's sleep I fell into the State
And I hunched in its belly till my wet fur froze.
Six miles from earth, loosed from its dream of life,
I woke to black flak and the nightmare fighters.
When I died they washed me out of the turret with a hose.

–Randall Jarrell, 1945

An Irish Airman Foresees His Death

I know that I shall meet my fate
Somewhere among the clouds above;
Those that I fight I do not hate,
Those that I guard I do not love;
My country is Kiltartan Cross,
My countrymen Kiltartan's poor,
No likely end could bring them loss
Or leave them happier than before.
Nor law, nor duty bade me fight,
Nor public men, nor cheering crowds,
A lonely impulse of delight
Drove to this tumult in the clouds;
I balanced all, brought all to mind,
The years to come seemed waste of breath,
A waste of breath the years behind
In balance with this life, this death.

–W. B. Yeats, 1922

The War in the Air

For a saving grace, we didn't see our dead,
Who rarely bothered coming home to die
But simply stayed away out there
In the clean war, the war in the air.

Seldom the ghosts came back bearing their tales
Of hitting the earth, the incompressible sea,
But stayed up there in the relative wind,
Shades fading in the mind,

Who had no graves but only epitaphs
Where never so many spoke for never so few:
Per ardua, said the partisans of Mars,
Per aspera, to the stars.

That was the good war, the war we won
As if there were no death, for goodness' sake,
With the help of the losers we left out there
In the air, in the empty air.

–Howard Nemerov, 1986

To Poets and Airmen

Thinkers and airmen–all such
Friends and pilots upon the edge
Of the skies of the future–much
You require a bullet's eye of courage
To fly through this age.

The paper brows are winged and helmeted,
The blind ankles bound to white road,
Streaming through a night of lead
Where cities explode.
Fates unload

Hatred burning, in small parcels,
Outrage against social lies,
Hearts breaking against stone refusals
Of men to show small mercies
To men. Now death replies
Releasing new, familiar devils.

And yet, before you throw away your childhood
With the lambs pasturing in flaxen hair,
To plunge into this iron war,
Remember for a flash the wild good
Drunkenness where
You abandoned future care,

And then forget. Become what
Things require. The expletive word.
The all-night-long screeching metal bird.
And all of time shut down in one shot
Of night, by a gun uttered.

For the joy that was is hidden under grass,
Shadows of hawks flicker over.
Buried in cellars is laughter that once was
Which the pick and shovel endeavor
Vainly to uncover;
Like a child buried when the raiders pass.

With axe and shovel men hunt among the bricks,
With lamps and water, for their soul
Of lilac in the city square; they hack with picks
Amongst the ruins for their love's goal,
As though a smile frozen at the North Pole
Might take pity on their tricks.

–Sir Stephen Spender
London, December 1940
Dedicated to Michael Jones in his life, and now in his memory

Two Poems of Flight-Sleep

I. CAMDEM TOWN

Army Air Corps, Flight Training, 1943

With this you trim it. Do it right and the thing'll fly
Itself. Now get up there and get those lazy-
eights down. A check-ride's coming at you
Next week.
I took off in the Stearman like stealing two hundred and
twenty horses
Of escape from the Air Corps.
The cold turned purple with the open
Cockpit, and the water behind me being
The East, dimmed out. I put the nose on the white sun
And trimmed the ship. The altimeter made me
At six thousand feet. We were stable: myself, the plane,
The earth everywhere
Small in its things with cold
But vast beneath. The needles on the panel
All locked together, and a banner like World War One
Tore at my head, streaming from my helmet in the wind.
I drew it down   down under the instruments
Down where the rudder pedals made small corrections
Better than my feet   down where I could ride on faith
And trim, the aircraft slightly cocked
But holding the West by a needle. I was in
Death's baby machine, that led to the fighters and the bombers,
But training, here in the lone purple,
For something else. I pulled down my helmet-flaps and droned
With flight-sleep. Near death
My watch stopped. I knew it, for I felt the Cadet
Barracks of Camden die like time, and "There's a war on"
Die, and no one could groan from the dark of the bottom
Bunk to his haggard instructor, I tried

I tried to do what you said   I tried   I tried
No;   never. No one ever lived to prove he thought he saw
An aircraft with no pilot showing: I would have to become
A legend, curled up out of sight with all the Western World
Coming at me under the floor-mat,
minute after minute, cold azures,
Small trains and war bound highways,
All entering flight-sleep. Nothing mattered but to rest in the winter
Sun beginning to go
Down early. My hands in my armpits,
I lay with my sheep-lined head
Next to the small air-moves
Of the rudder pedals, dreaming of letting go   letting go
The cold   the war   the Cadet Program and my peanut-faced
Instructor and his maps. No maps   no world   no love
But this. Nothing can fail when you go below
The instruments. Wait till the moon.   Then.   Then.
But no. When the waters of Camden Town died, then so
Did I, for good. I got up bitterly, bitter to be
Controlling, re-entering the fast colds
Of my scarf, and put my hands and feet where the plane was made
For them. My goggles blazed with darkness as I turned,
And the compass was wrenched from its dream
Of all the West. From luxurious
Death in uncaring I swung
East, and the deaths and nightmares
And the training of many.

—*James Dickey*

The Firebombing

(ABBREVIATED VERSION)

Denke daran, dass nach den grossen Zerstörungen
Jedermann beweisen wird, dass er unschuldig war. –Günter Eich

Or hast thou an arm like God? –The Book of Job

Homeowners unite.

All families lie together, though some are burned alive.
The others try to feel
For them. Some can, it is often said…

Starve and takeoff…

Snap, a bulb is tricked on in the cockpit

And some technical-minded stranger with my hands
Is sitting in a glass treasure-hole of blue light,
Having potential fire under the undeodorized arms
Of his wings, on thin bomb-shackles,
The "tear-drop-shaped" 300-gallon drop-tanks
Filled with napalm and gasoline.

Thinking forward ten minutes…

There is then this re-entry
Into cloud, for the engines to ponder their sound.
In white dark the aircraft shrinks; Japan…

Enemy rivers and trees
Sliding off me like snakeskin,
Strips of vapor spooled from the wingtips…
Sunday night in the enemy's country   absolute…

Going: going with it…

Rivers circling behind me around
Come to the fore, and bring
A town with everyone darkened.
Five thousand people are sleeping off
An all-day American drone…

In a dark dream that this is
That is like flying inside someone's head…

I still have charge–secret charge–
Of the fire developed to cling…

Not atoms, these, but glue inspired
By love of country to burn,
The apotheosis of gelatin…

On altitude, drones on, far under the engines,…

In sleep, as my hand turns whiter
Than ever, clutching the toggle --
The ship shakes bucks
Fire hangs not yet fire
In the air above Beppu.
For I am fulfilling

And "anti-morale" raid upon it.
All leashes of dogs
Break under the first bomb, …
Their heads come up with a roar
Of Chicago fire:…

In a red costly blast
Flinging jelly over the walls
As in a chemical war-

Fare field demonstration.
With fire of mine like a cat…

Fire shuttles from pond to pond
In every direction, till hundreds flash with one death…

The death of children is ponds
Shutter-flashing; responding mirrors; it climbs
The terraces of hills
Smaller and smaller, a mote of red dust
At a hundred feet: at a hundred and one it goes out.
That is what should have got in
To my eye…

One is cool and enthralled in the cockpit,
Turned blue by the power of beauty,
In a pale treasure-hole of soft light
Deep in aesthetic contemplation,
Seeing the ponds catch fire
And cast it through ring after ring
Of land: O death in the middle
Of acres of inch deep water! Useless…

In this detachment,
The honored aesthetic evil,
The greatest sense of power in one's life,
That must be shed in bars, or by whatever
Means, by starvation…

O then I knock it off
And turn for home over the black complex thread
worked through
The silver night-sea,
Following the huge, moon-washed steppingstones
Of the Ryukyus south,

All this, and I am still hungry,
Still twenty years overweight, still unable
To get down there or see
What really happened…

–James Dickey

The Liberator Explodes

There, in the order of traffic
Of aircraft. Where one of them once
Was moving, in a clumsy hover,
It is like a blow through the sky
That does not move.

Why would you watch it
Before it becomes of fire?
There are many arranged on the air.
This one you might be watching,
Held in fear

That contains no fear, but boredom, or fascination,
As it turns on the final approach.
Or you might be watching another
That does not fall.

If it is this one, you see
For an instant, nothing special. It is hanging down
As it would, the big wheels not spinning,
And now are fire:

One shot, a great one,
By accident takes place where the plane is:
The plane was. All of it is gone
Save the part that goes in on one wing,
There, off the end of the runway.

Then comes the shape
Of a silence made of an army
In one breath all watching wildly.
Things move out, and toward
Where it must have come down

There, off the end of the runway,
Still alive with a little of fire.
Here is the purest of fact
That took place like the purest of symbols.
The mind fires over and over

An aircraft that has blown away distance,
But cannot fetch that fact,
Or remember or know or imagine
What the faces of those must have felt
There in the brief shot of light,

And so must lie down again, and again,
Below the ground moved by palm-leaves
Of the mind of that time, and let that fade,
And lie in the luck of salvation
In the cities,
In the suburbs of time, until

There cracks across the simplest of the mind's
Eyes, that purchase of terror on the air,
The burst of light within flame,

Magnificent, final, and you behold your own
Unmirrored face freely explode,
And face, beyond faces,
Your brother of parallel fire.

–James Dickey

First Snow on an Airfield

A window's length beyond the Pleiades
Wintering Perseus grounds his bow on haze
And midnight thickens on the fall of snow.
Now on the sound of sleepers past their days
The barracks turns to myth, and none shall die
But widen and grow beautiful a while
And then be written on a Grecian sky.

Look, the burnt mountain whitens, and the trees
Grow cavernous. And the field's lights are spread
Spangling on the daubed and rushing air
That fills with drone of engines overhead.
And see: the constellations of the running-lights,
Crossed on the beacon's arm, bring home the planes
That almost layered the hills with trilobites.

As near as a chance: A winter's memory
Of seconds not too soon that might have been
Fossils at impact with the shrouded stone
Here on the ground, the noise of a machine
Above the falling snow at season's turn
Memory crossed with moment – and again.
Tomorrow's manual of guns to learn.

–John Ciardi

P-51

It fills the sky like wind made visible
And given voice like drums through amplifiers,
Too great a terror to be lost on death
Remembering that all our dreams are fliers.

This terror, cannoned as the hawk is billed,
Taloned with lusty boys who love their toy,
Mounts on the living energy of grace
Whose passing cracks on burning lathes of joy.

Piston by piston the made fumes of flight
Frenzy the startled air her passing sears.
Fast as a head can turn from East to West
She summons distances and disappears.

That moment only – glancing up and gone –
And see, her boy outburns the burning year.
And we are clod and pasture fixed upon
Her birth above the hills like a crowd's cheer.

–John Ciardi, 1945

Return

Once more the searchlights beckon from the night
The homing drone of bombers. One by one
They strike like neon down the plastic dome
Of darkness palaced on our sea and sight
Where avenues of light flower on a stone
To bring the theorem and its thunder home.

Wheels touch and snub, and on the wing's decline
From air and motion into mass and weight
Grace falls from metal like a dancer's glove
Dropped from the hand. She pauses for the sign
Of one more colored light, and home and late
Crosses to darkness like an end of love.

Under the celebration of the sky
Still calling home the living to their pause
The hatches spill the lucky and returned
Onto the solid stone of not-to-die
And see their eyes are lenses and they house
Reel after reel of how a city burned.

–John Ciardi

The Pilot in the Jungle

I.

Machine stitched rivets ravel on a tree
Whose name he does not know. Left in the sky,
He dangles from a silken cumulus
(Stork's bundle upside down
On the delivering wind) and sees unborn
Incredible jungles of the lizard's eye:
Dark fern, dark river, a shale coliseum
Mountained above one smudgepot in the trees
That was his surreal rug on metered skies
And slid afire into this fourth dimension
Whose infinite point of meeting parallels
He marks in ultra-space, suspended from
The chords of fifty centuries
Descending to their past – a ripping sound

That snags him limb by limb. He tears and falls
Louder than any fruit dropped from the trees,
And finds himself in mud on hands and knees.

II.

The opened buckle frees him from his times.
He walks three paces dressed in dripping fleece
And tears it off. The great bird of his chute
Flaps in the trees: he salvages its hide
And starts a civilization. He has a blade,
Seventeen matches, his sheepskin, and his wits.
Spaceman Crusoe at the wreck of time,
He ponders unseen footprints of his fear.
No-eyes watch his nothing deep in nowhere.

III.

He finds the wreck (the embers of himself)
Salvages bits of metal, bakelite, glass–
Dials twisted from himself, his poverty.
Three hours from time still ticking on his wrist
The spinning bobbins of the time machine
Jam on an afternoon of Genesis
And flights of birds blow by like calendars
From void to void. Did worlds die or did he?
He studies twisted props of disbelief
Wondering what ruin to touch. He counts his change
("Steady now, steady...") flips heads or tails and sees
The coin fall into roots. An omen? ("Steady...")
He laughs (a nerve's slow tangling like a vine)
Speaks to himself, shouts, listens, hears a surf
Of echo rolling back to strand him there
In tide pools of dead time by caves of fear,
And enters to himself, denned in his loss,

Tick-tick, a bloodbeat building on his wrist.
Racheting down the dead teeth of a skull
(The fossil of himself) sucked out of sight
Past heads and tails, past vertebrae and gill
To bedrocks out of time, with time to kill.

–*John Ciardi*

Visibility Zero

All day with mist against the hurdling wind
The lights hung dressed in halves and a blur.
Air that was solid on a hurtling wing
Hangs sodden, and the parked planes wear like fur
Their look of waiting in the liquid pause
Of cloud descended, in a veil of gauze
The three complete and only trees incite
Their separate loss into the early night.

Fixed to the gauge that swears we cannot see,
Our engines, blind as junk, await the light.
Cards, dice, and spinning coins turn noisily
Into the separate corners of the night.
This was the day we saw our lives made safe,
The day no engines burned and no one gave
A morning thought to chance, but late in bed
Praised the tiered fog that nowhere touched the dead.

Complete in pause, we woke into no need,
Turned back to sleep, stayed dry, and wished for mail.
Ate, and addressed a holiday – a nod
To cancelled schedules, and a word to tell
Our postponed fear that it was not our choice.
And then, released, the barracks lounging voice

In praise of hours when instruments agree
We need not waken and we need not see.

–John Ciardi

Darius Greene and His Flying-Machine

(ABBREVIATED VERSION)

If ever there lived a Yankee lad,
Wise or otherwise, good or bad,
Who, seeing the birds fly, didn't jump
With flapping arms from stake or stump,
Or, spreading the tail
Of his coat for a sail,
Take a soaring leap from post or rail,
And wonder why
He couldn't fly,
And flap and flutter and wish and try–
If ever you knew a country dunce
Who didn't try that as often as once,
All I can say is, that's a sign
He never would do for a hero of mine.
An aspiring genius was D. Green;
The son of a farmer, age fourteen;
His body was long and lank and lean–
Just right for flying, as will be seen;
He had two eyes as bright as a bean,
And a freckled nose that grew between,
A little awry – for I must mention
That be had riveted his attention
Upon his wonderful invention,
Darius was clearly of the opinion

That the air is also man's dominion,
And if you doubt it,
Hear how Darius reasoned about it.
"The birds can fly an' why can't I?

–John Townsend Trowbridge, 1869

Flight

Solitude, Freedom, Beauty, Mystery, Meaning, and Motivation

Wings of My Silver Plane

Grazing the broad blue skylight
Lip where the falcons fare,
Riding the realms of twilight
Brushed by a comet's lair.
Snug in my coat of leather,
Watching the skyline swing,
Shedding the world like a feather
From the tip of a tilted wing.
There are trials that I can travel
When the years of my life wane
But I'll let a little rainbow ravel
Through the wings of my silver plane.

–Author Unknown

A Distant Thunder

We climb westward.
Our breathing and the noises of flight
mar the stillness
as I guide our fighter's awesome might.

The sun is alive.
Its glaring orb spills over my gunsight
and into my eyes.
It burns into my brain, leering and bright.

The sun is dying.
Its earth angle is now very slight.
Darkness veils the plains.
The mountains alone are robed in white.

The sun is dead.
The flaming purples and pinks of twilight
have given way
to the enchanting charcoal black of night.

We are immersed.
Our instruments bathe us in soft red light.
The earthbound hear us pass;
their longing minds propel them to our height.

–Dallas Blevins

Those Who Fly

I fly above the distant world
the beauty of life below, unsurled.
The clean crisp air and blessed earth
was all God's plan for us at birth.
Bathed in sun, a clear days flight
melds into darkness, twinkling stars of night.

The sharpness of trees and colors of day
streetlights and bright lights of night give way
to beauty of moon following the wing.
Each day, each night, each flight, does bring
a peace and joy known by no other
than those who fly, bonded, as sister and brother.

–Patricia Rockwell, 1992

Free

I see everything.
I am in complete control.
The horizon lined in the shadows of the setting sun.
The brilliant burning blue beckons me.
The clouds dance relentlessly with the flirting wind,
leading me to the heavens.
The crisp, cool air cleanses my mind and soul.
I am free now, if just for a moment.
Free from the hypnotic cadence of the world below.
Free by definition of the sky.

–Victoria Schrauwen

A Dance Through the Sky

I dance through the sky on shining red wings,
I loop and I roll and do other such things;
And when I'm up here I have not a care,
for here I belong I'm at peace in the air;
The blue and the green of the earth dance before me,
as good friends look up to see me performing;
This dance that I do for the people to see,
takes lots of hard work but it's all fun to me;
So when I come down just ask me to dance,
we'll waltz through the clouds if you just take the chance.

–Timothy S. Bastian, 1997

Why I Fly

Oh when I fly,
up in the sky,
upon the wings of man;
It feels so great,
it's like a gate,
in to a wonder land;
And when I fly oh so high,
among the eagles grace,
I see the things that can't be seen,
in any other place;
And when I lift up off the ground,
and break the bonds that hold me down,
I just feel so very free,
like the eagle over me;
And if you still do not know why,
just spread your wings and give a try,
you just may learn the reason why,
it is that I love to fly.

–Timothy S. Bastian

Gift

The scenario set clearly,
I steal toward the airplane,
Relishing damp grass, expectant shivers,
A cold, cloudless sky and
Hint of yellow-white eastward where
The sun rushes up
To dazzle this day into being.
A sensuous suite of sensations, this;
Trembling calm,

Knowing this machine will take me
From this hard ground to sparkling space where
no-one, Save God Himself
Can have dominion over me.
Airborne, always better
Than its dreams,
I rise to the sun
At the edge of the world
In the icy sky which lifts with loving hands;
Washes me clean of earthbound dust.

Rapt above joy,
Elemental belonging,
For now, apart;
Orphaned as angels
Between death and heaven;
Intemporal, alien-
Sifted by the Master's hand.
Mountains, rocky ripples now,
Topped and scaled in giddy climbs and glides.
Dank, patchwork prairie shyly shifts
Toward a warming sun.
For now, no thought of return,
Held in space alone, careening,
Screenplay for tomorrow's dreams.

–*James MacNutt*

On Top

Resonant,
A bass and tenor chorus
In sublime harmony,
The engines pacify me.
In this pristine palace
Over a cotton landscape
With no world to see,
Only sky,
Horizonless heartland
Of near-space,
Keening,
Apart, we sail.
And here,
All the voices of God
Become distinct.

—James MacNutt

Sunrise Flight

In morning sun I've threaded canyons,
Buttes and cliffs in slanted rays,
Riding currents, topping crests, and
Sinking down toward valley floors.
My elbows are her wingtips, flexing,
Face a flush in spinning fury;
Blades a shining disk to sunward,
Flickering now, in passing shade.
My heart, her engines breathing freely,
Racing happily aloft.
Tall spruce reaching upward, straining,
Clouds above, and we between, in

Timeless, spaceless grace, cavorting,
Reaching all directions.
In morning sun I've leapt from earth
To sweep through canyons,
Buttes and valleys;
Dancing on the winds of Terra,
Poised on metal wings.

–James MacNutt

The Surpassing Way of the Sky

Those who dare to venture into an unexplored land will have
revealed to them things which were never known.
Those who venture out upon the sea will have revealed to
them things never heard.
But those who venture into the sky upon wings of silence...
Yes, the ethereal adventurers...
Theirs is the revelation of things never even dreamed!
Such are the ways of the explorers
And the surpassing way of the sky.

–Gary Osoba, August 1999

Evening Flight

The wavelength stretches. Blues and greens
shoot into space. Refracted reds and yellows
bathe the mountainside below. Its mellow tones,
breath catching; as I turn and ride down light,
fall into the shadows on the eastern slope,
and then fly outwards, seeking the sunlit plain.

Only the shade line moves. The air is still.
The smoke slow drifting in a wayward plume,
where earthfast houses seek to drive away the cold,
whilst I surge upwards, from near-dark, into light;
rejoice in icy nip of fingers, and of face,
escape the eastward roll of night-time down below.

This is the pilot's moment. Stretching out
the ending of a day (that packed into its span
those blustery take offs, and the working ridge)
until, from high, the sinking of the sun into the sea
compels a gently wheeling, slow descent
to high speed touch down, in the field of night.

–David Pedlow

U2

(FORMAT MODIFIED FROM THE ORIGINAL POEM)

I am the eye of God.

I live, high above the upper crusts of atmosphere. Inside my jaws I withhold a man – who calls himself pilot and dreams of astronauts. Who, beneath his body, beholds all creation with a glance. I am the abomination of flight.

The perfection of all desire for escape and greed for curiosity: to see what cannot be. I can see a child scream, the button on a tweedy jacket, or two hands cross from twelve miles high. The Earth, a cross-sectioning of infra-red and laser targeting on tangents, lies below me for my pleasure. The ugly surface of the planet revolves around my frame.

Here, between ionosphere and troposphere, my trailing wings, a color-less black, cast no shadow on a cloud. Here the compass rose is cast aside. I have no need for compass magnet or direction. The sun's above, then passing under. Rises by falling from humanity.

I am the abomination of God. In flight, I aim along a camera sight: the particularities of death in life. I have a purpose.

The Earth must stay exactly as it is.

–P. H. Liotta

Feeling Compassionate

Feeling compassionate,
you deliver me into the arms of the clouds,
bundled and new-born.
With a swift and sunlit thrust,
the dismal earth disappears,
and all signs of small and dutiful humankind
are buried in cushioned, white silence.

Up here, time stands still,
order is gentle and gliding.
Up here, the center is certain
as I hold onto you.
In one suspended moment of calm,
the truth of sunlight,
reflecting off the tops of billowing clouds
and into your eyes, is captured.
We could live here forever, with wind and eagles,
kisses and sighs.

The moment hovers,
lasting not long enough,
until we both, unspoken,
consent we must descend,

break through this Heaven
and return to the underside of what is
all too lovely.

Your lips move into the small microphone,
connecting wires, transmitting the OK
to reenter the other reality.
I wish those lips to be reserved
for blue sky, clouds, and me.

But they are not,
and we break through.
You try so hard to let me down easy.
Still, the landing jolts
and the gravity of gray earth pulls feet to ground,
pulls me away from you.

So that opening the door to the blue kitchen
with dishes waiting and "have-to's"
magnetized on the refrigerator door,
my center is shivering,
knowing it is left somewhere up there
in the perfectly shifting clouds.

–Carolyn Berge

Dawn

Swinging down the sky blue trail
Early in the mornin',
Earth all veiled in purple mists
At the hour o' dawnin',
Silver streaks across the blue,
Casting down their shadows

On the mists that swim beneath
Like Olympian Meadows.
Swinging 'bove the sky blue trail
Mornin' stars are shinin',
Moon gone pale with breaking light
Like lover pinin';
Silence broods across the deep,
Mystic voices only
Whisper through the azure dawn
To the watcher lonely.

Lying 'neath the sky blue trail
Rivers go abendin',
Seeking each its heart's desire
In the oceans endin',
Silver ribbons that will lead
Who has heart to follow,
Far beyond the fields that lie
Crowned with harvest fallow.

Swinging down the sky blue trail
To feel the joy o' livin',
Drinking till your soul is filled
With the wine o' heaven;
Who would barter days on earth
For one hour o' dawnin',
Swinging down the sky blue trail
Early in the mornin'.

–Gill Robb Wilson

Escape!

It's been many decades since I first flew a falcon from my fist.
Many decades since I first tasted flight in a Rogallo-type design.
Many decades since my first soaring experience,
first hour-long flight, and first cross-country excursion.
Be that as it may, this is also true:
Yesterday, deep in the throes of winter and while trapped in traffic,
my eye was caught by a glimmer of light
leaving the wing of a gull
As I watched him climb,
he turned ever so precisely on the horns of the thermal...
nearly stopping in the moment.
My heart first skipped a beat, then sped up, then began to race.
The bars on the cell of time cannot hold these things captive.
They are unchangeable.

–Gary Osoba, January 1999

Pilots

CHARISMA, PERSONALITY, AND LEADERSHIP

Because I Fly

Because I fly
I laugh more than other men
I look up
And see more than they.
I know how clouds feel
What it's like to have the blue
In my lap.
To look down
On birds
To feel freedom in a thing called the stick
Who but I
Can slice between God's billow-legs
And feel them laugh and crash with His step?
Who else has seen the unclimbed peaks?
The rainbow's secret?
The real reason birds sing?
Because I fly
I envy no man on earth.

–Author Unknown

To Pilot

To pilot . . . to experience a rare ecstasy known only to
the few who have coaxed an airplane to uplift and to fly.

To gain a mounting prowess; to conduct and relish
carefree journeyings above unknown, inviting,
exquisite earthly mosaics.

To arrow majestically and alone, to venture
touchdowns at heavens far beyond the home airfield.

To feel body and mind tingle amid the humbling
stimuli of nocturnal flight, when great city and isolated
farm shimmer and twinkle their greetings,
and diamond stars reply.

To sigh in gratitude as the spirit ascends, unfettered
in the vast uncluttered concave of the sky.

To savor the pure keen enchantment of flying across
the broad castellated heavens.

To recognize that air's global oceans can roil with a
prodigiousness sufficient to ground the boldest aviator;
for the air which frees us is never to be taken lightly.

To unabashedly extol air piloting; to consciously
promote this joyous, skyward pursuit, that there may be
many more to reap its grand beneficence.
Ah, to pilot...

–Desmond M. Chorley

No Stalling

2300 rpms
95 knots
cruising speed

life is not like that
nor should it be.......for long.
trying to find poetry
for a new experience
that thrills me
and scares me..
and scared ...I want to be.

and thrilled... I need to be.
no parachute.
It's the only way.
too old for less
more old?
more willing.

–Dave B. Nichols

To All Aircrew

There is a bond between them
That only they can share,
Whose lives are bound together
By the friendship of the air.
At home in any company,
No matter where they are,
From Singapore to London,
From Cyprus to Accra.
No petty rules prevent them

Relaxing as they wish,
In backstreet bars OR ballrooms,
No trace of snobbishness.
A classless sort of people
With backgrounds far apart,
Born of Lords AND miners,
With flying in their heart.
But when they're not relaxing
That is a different case.
The rules are hard and rigid
And there is no easy pace,
Or room allowed for error,
In decisions that they make.
No second chance is given
With so many lives at stake.
The public thinks it's easy
And say they're over paid,
Complain of noise and nuisance
Each single flight that's made.
Could they but see the lightning flash
Amidst the monsoon rain,
Know half the problems to be faced,
Perhaps they would think again.
Most of them are married
With children like your own
And do not relish nights away,
Their families left alone.
It isn't all wine and roses
Although it may appear as such.
Just folks who know how to live
And love to live so much.

–Doug Atkins, 1960s

The Aviator Man

Day after Day
He was smooth and cool
Never afraid
To break a rule

He was running the race
and climbing the hill
Conquering his fears
and feeling the thrill

He was on his way up
And moving fast
Never doubting
That it would last

A boy, a child, a man
Who made himself a name
He reached for his dreams
And finally found his fame

But, in one simple moment
This picture did fade
All was quiet
And a legend was made

He sailed the skies...
He followed his plan...
He earned his wings...
For he was the AVIATOR MAN

–Clay Greager

To a Pilot's Wife

His first love is the plane he flies
day after day across God's skies;
she is his 'mistress of the blue'
whom he adores as much as you.
But she is just a passing thing
who does not share his wedding ring;
so do not envy her, my friend,
this other love someday will end.
Your strength lies in the vows you made
when you were young and unafraid,
and also in your children's charms–
from his fulfillment in your arms.
Someday your Pilot will retire
to sit with you before the fire;
then *both of you can scan the skies*
for your young son with Pilot's eyes.

–Gene Griener

The Man in the Cabin

Here's how! To the men in the cabin!
The top hands of people who fly
The lads who must hit on the button
The cone that is set in the sky.
The man in the cabin must know his ship,
Each longeron, cable and baffle and clip,
Each valve to the fuel and pitch to the prop,

Where temperatures lessen and pressures must stop,
Each instrument principle, function and make,
Each meter jet, rudder tab, slinger ring, brake,
What octanes are proper, how mixture should change,
What manifold pressures are safe at what range,
What air speed is sloughed when the wheels go down,
The set of the flaps when approaching the ground,
And nothing that's utilized out of the run
But they in the cabin know how it is done.

The man in the cabin must know the land,
Reaction to heat of the rock or the sand,
Refraction of light on the colors below,
The greens and the yellows, the white of the snow,
How contours will cause him a turbulent flight,
How eddies will cease with the falling of night,
The headlands that give off a multiple course,
The dust storms, velocity, movement and force,
All airports and highways and railroads and mills,
The smoke o'er the cities and haze o'er the hills,
And nothing in all of the sweeping terrain
But what to the man in the cabin is plain.

The man in the cabin must know the sky
With the range of its moods and the reasons why
He can ice up a wing with a load of sleet
When the temperature change and a dew point meet,
He must judge of the rise of the thunderhead
And the static discharge of the light'ning shed,
The inversions and pressures and highs and lows,
How the cold front forms and the path it goes,
The habits of fog from the land or the sea,
At what altitudes seasonal winds will be,
And there's nothing that broods in the fickle skies
But the man in the cabin must analyze.

Then here's to the men in the cabin!
The top hands of people who fly,
The uniformed force of the airways
Who wangle their chuck in the sky.

–Gill Robb Wilson

The Old Helmet

This ragged old leather is beaten and grey
And though helmets are only for pictures today
I keep the worn relic, in faith or in fear
That it's more than a helmet . . . this worn souvenir
Tucked safe in the crown of my uniform cap
While I boot the big cargo ships over the map.

I first wore the helmet while learning to fly
As an air corps cadet in a war crazy sky
Where rain, sleet and snow and my sweat gave it mould
Under hot Texan suns and the bleak Texan cold:
It saw me through solo, cross-country and spins
While I learned how a soldier who wants to cry, grins.

In Jennies and Nieuports and Bristols and Spads
And the war profit junk which they furnished us lads
It blotted up castor oil, gas fumes and dust,
Became powder burned, blood soaked and colored like rust,
Was lost 'tween the lines but was found by the crew
Which had dragged me safe in when the Jerry got through.

When armistice came I returned to the sticks
Where we barnstormed the pastures and stunted the hicks
In an old crate through the highland and low
We got ten bucks a ride and we took in the dough,

The strapless old helmet the symbol of things
Which a man learns to value who lives by his wings.

The airmail was just a fresh war to be won
Where we died to prove blind flying could not be done
A compass strapped onto one's knee for a guide
With a ship that resembled a brick in a glide
And weather maps made once a week for the grange
And for chaps in the sheep wagons out on the range.

But some of us lived to see science unfold
In the turn-and-bank, air speed, controlled manifold,
The Sperry, the super-charge gear and the prop
With the Kollsman, de-icers and beam for the top,
The beacons and range finders, transmitters, speed
And with all the flight plans and controls that we need.

My co-pilots grin when they look in my cap
As I lay it aside for a smoke and a nap,
They know the old man is in need of his rest
And the weather looks good down the beam to the west
But when it takes faith in addition to lore
The old helmet's back in the money once more.

–Gill Robb Wilson

Cloud Dreamers

A personal revelation
Few people know what makes us tick
Those few of us who fly till' we're sick
We do it for fun, we do it for kicks
Spilling through clouds, learning new tricks
This one selfish act, we perform next to God

Enjoying being lonely, thinking...
This pleasure of mine I share with a few
Those who stick to the skies and prefer a dark brew
Keeps us close to our thoughts, allows us to roam
We dream of that feeling that brings you back home.

–Patrick Hamilton

The Aviator

I am not a Charles Yeager,
And I'm sure not Richard Bong;
In fact most things that I attempt
In airplanes turn out wrong!

But I love the smell of ADI,
And engines that are round,
And I love to hear the big-bore bird,
With it's deep, un-bridled sound.

And I love to go to Reno,
And I love to watch them fly,
Tho it breaks my heart to see
A fellow aviator die.

But I know they'll go on flying,
And I know that so will I,
For nothing short of death will ever
Keep us from the sky.

And I'll never know what makes us so,
Were we just born to fly?
Or did we get hooked the first time we saw
An airplane in the sky?

Now some men feel a three piece suit
Is proof they've passed the test,
Or driving a Mercedes Benz,
Or a Rolex on their wrist.

I shall be forever grateful,
That He chose-me from the rest,
To spend my life in uniform,
And brother to the best.

And when my life is over,
And it's time that I should die;
I hope He'll let me join the crowd
In His hangar in the sky.

And there, we'll raise another glass
And re-tell all those lies,
About our mis-adventures
Way up yonder, in the skies.

–Michael J. Larkin

Charles Yeager is a test pilot notorious for breaking the speed of sound. Major Richard Bong (1920–1945) was the United States' most successful fighter ace in World War II shooting down 40 Japanese aircraft. Major Bong was awarded the Medal of Honor by General MacArthur.

"Come and Fly"

One day I will fly,
on wings of eagles I'll follow my mentors in yonder sky –
Since childhood I've honored,
Like knights of Camelot, full of dare,
Those brave and daring souls who walk upon the air –

With child-like awe I've looked into their eyes wanting to know,
hoping to embrace their spirit, looking for sign
in the direction to go –
Now as a man, I stand alone listening–
I hear a thousand distant voices calling from the firmament,
So embedded is this yearning, so strong the beckoning,
....."Come follow where our footsteps went."
"Come and fly, come and fly," "Oh, young fellow come and fly."
I long to be a part of the brotherhood who dance across the sky,
I yearn to follow in their footsteps,
hoping one day I'll fly. –

–David A. Morris

Ode to My Hero

For years I watched you airbrush contrails against
the vail of endless blue,
In admiration from airport terminals I followed your footsteps
wishing I could be just like you. –
For years I watched you traverse the sky's clouded wake, just
soaring by,
Waving as you went out of sight, wishing that I could fly. –
I often met you face to face but didn't quite know what to say,
for fear of certain embarrassment, I quietly went on my way. –
I want so much to be like you, to have that common ground we'll
share,
To be windswept along side my hero, with the whisper of the air. –
This is my life-long wish my heroic friend,
To carry on the centuries old tradition
of the brotherhood, to ensure
that it never comes to an end. –

–David A. Morris

Untitled

Someday we will know, where the pilots go
When their work on earth is through.
Where the air is clean, and the engines gleam,
And the skies are always blue.
They have flown alone, with the engine's moan,
As they sweat the great beyond,
And they take delight, at the awesome sight
of the world spread far and yon.
Yet not alone, for above the moan,
when the earth is out of sight,
As they make their stand, He takes their hand,
and guides them through the night.
How near to God are these men of sod,
Who step near death's last door?
Oh, these men are real, not made of steel,
But He knows who goes before.
And how they live, and love and are beloved,
But their love is most for air.
And with death about, they will still fly out,
And leave their troubles there.
He knows these things, of men with wings,
And He knows they are surely true.
And He will give a hand, to such a man
'Cause He's a pilot too.

–Author Unknown

Buckaroo

The Buckaroo rambles the prairies of blue
Where the cloud is the chapparal patch,
His home is the range where temperature change

Is the red skin of pony dispatch:
A cayuse of steel from a maverick herd
And a loop made of sound for a rope
To corral the hills where the beacons are set
For the buckaroo's pony to lope.

His drink is the dew of the thunderhead crag
And his meat is the wild eagle breast,
The star dust is alkali white on his face
While the meteors tattoo his chest:
A sound in the sun and a flame in the moon
Is the spoor of the trail that he rides
And the mist of the rain on the storm trampled plain
Is the cave where the buckaroo hides.
The dawn is his gift from the day and the night
Which they wove out of purple and gold,
The cool forest sheen is his saddlecloth green
With the white rivers lining the fold:
The dusk is his gift from night and the day
As a canvas to spread for his camp,
The song of the wind for the peace of his heart
And the light of the moon for his lamp.

And when his remuda is broken and worn
And the last cayuse stumbles and falls
To bleach there alone on the prairies of time
Where the voice of the timber wolf calls,
His soul will still ramble the high open range
As the master of sinners has planned,
The mercy of God for the beam that he rides
On a flight plan he now understands.

–Gill Robb Wilson

The Last Bouquet

I've flown 'em all from then till now
The big ones and the small
I've looped and zoomed and dove and spun
And climbed 'em to a stall,
I've flown 'em into wind and storm,
Through thunder clouds and rain
And thrilled the folks who watched me roll
My wheels along their train.
I've chased the steers across the range,
The geese from off the bay,
I've flown between the Princeton towers
When Harvard came to play,
I've clipped the wires from public poles
The blossoms from the trees
And scared my best friends half to death
With stunts far worse than these.
The rules and codes and zones they form
Are not for such as I,
Who like the great wild eagles fling
My challenge to the sky,
A bold, free spirit charging fierce
Across the fallow land –
And don't you like these nice white flowers
I'm holding in my hand?

–Gill Robb Wilson

Listen, Sister!

The Infantry and Engineers
Have gallant men in ranks
And so have Coast Artillery
And Cavalry and Tanks
But when the Air Corp's noble sons
Parade across the sky
The hearts of all fair maidens thrill
To see them passing by.

At Langley, Wright and San Antone
They bring the squadrons down
And get Marines to guard the ships
And then they go to town
And sister when you hear them come
You better hunt your ma
For I've been in the Air Corps
And I'll tell you what I saw.

I saw a lass at Mitchell Field
Forget that she was wed,
Forget her past, forget herself
Forget what people said:
Forget she led the village choir,
Forget she hoped to be
The Garden City Candidate
For charm and chastity.

I saw a lass at old Orleans,
Where all the maids are fair,
Forget to rouge her lovely lips
And comb her raven hair,
Forget to pine, forget to sigh,
Forget to lock the door

Because the boys from Barksdale Field
Had landed there before.

I heard Mount Clemens fairest dream
Who always answered "no,"
So cold that when she went outdoors
The rain would turn to snow,
Get hot and tell a pilot "yes"
And all her scruples yield
And follow him because he flew
The ships at Selfridge Field.

I heard a maid from Sunnyvale
By Frisco's Golden Gate,
Admit she left an Admiral
In his retired estate
And whisper that she'd fly with me
To Jersey's sunny shore
Because she couldn't live without
The Aviation Corps.

So all the land is full of maids
North east and south and west
Who have their choice of service men
But love the Air Corps best:
Not Navy men nor Cavalry
Nor long haired Engineers
But Air Corps men who satisfy
The sweet and lovely dears.

And when their sons grow up to men
They'll join the Air Corps too
And have Marines to watch their ships
Just like their daddies do
And all the other service men
Will lift up grateful eyes

To hail the Air Corps and its sons
Who guard them from the skies.

–Gill Robb Wilson

PILOTS' RELATIONSHIP WITH THEIR AIRCRAFT

A Platform and a Passion

Happiness is brake release.
The roll is freedom in motion.
The oppressive elements of earth are broken with back pressure . . .
We are alone.

Every aircraft is the personification of femininity.
I entertain my aircraft like a gentleman;
She is truly fickle.
Therefore, it is only a friendship.

She is agreeable and unobtrusive.
When she is recalcitrant and does not respond, I am indignant.
If I go elsewhere, when I return she is cold and indifferent. . .
Profanity comes easily.

She has an always-forgiving, extremely gregarious disposition.
We have much in common.
We make up with intensity what
our association lacks for duration. . .
We live fast.

I am never guilty of irrational reverence,
The obsessive devotion breeds contempt.
Careless men lose self-respect.
Without self-respect, one is capable of abuse only.

Our relationship is not forced but natural.
When it is necessary to depart, there is reluctance.
I thank her and comment that I will call again.
And if I do not, I will speak kindly of her.

I am a pilot, I am proud.
What I fly does make a difference.
But regardless of my lot, I am neither ashamed, nor apologetic, nor envious.
For I have that face-down-trump, runway-behind-me, to-hell-with-the-world feeling.

I will probably never see the enemy face to face,
But I know that he too has a fashionable love.
I know that he enjoys the sensuousness of flight.
Such a love is universal.

–Ronald E. Pedro

The Seat

The Seat
It's ugly and worn,
lopsided and torn.
It's lumpy and wet,
from coffee and sweat.
"It reeks so bad!", they all say,

from no APU with a nine leg day.
It's too hard when you're too soft,
remember those hours you spent in LOFT?
But it calls you back, it knows your name,
whispers "the view from here, it's just not the same".
With all of this, it still can't be beat,
you're PIC now, so strap in the seat!

–Richard L. Barlow

To All Pilots

I sat upon a cold stone steed,
steel black bit in hardened mouth.
My spurs sink in but steel doesn't bleed,
I turn my head to the wind from the south.
But fools cannot fly, cast nostrils don't snort
nor whinny nor howl. I sit there 'til night.
The reins held loosely, I hold my court
with dreams, until dawn dies with the light.
You who put wings on steel spun sides
and wires and strings between your thighs.
A nudge with a knee, mouth comes into life,
giving some rein, your steel steed flies!
Fools, those blown by the wind, they have no wings.
What are their dreams, if life is only birth?
Your vestal white steed's propeller mane brings
and breathes life from air, water, fire, and earth.

–Patty Wagstaff

The More True Love

Infinite heavens
deeper than her temporary love,
dark blue shade
more real than her fickle eyes.
Unchanging expanse
bears me aloft on my metal wings.
Unpredictable lightning tempest
more welcome than her indecisiveness.
Wonderful smells constantly abounding:
her clothes, perfume and hair?
No, the intoxicating
smell of the earth and air.
The sky's liberating embrace,
comfortable, hot, or cold,
more welcome than her half-hearted hugs
and all the platitudes told.
Confessed to her all I was
and all I hoped to be,
all the promises of love
simply thrown back at me.
In that dark hour,
when I've had all I can bear,
I look up to the evening sunset,
at the J-3 dancing there.
When I feel a Wright thunder past,
or hear a Pratt whine by,
I know I'll always have a love
that is the bright blue CAVU sky.

–H. Gene Johnson

Training and Solo

The First Time

"I have a few questions...", I heard him say,
As my mind began to drift away,
To manuals, flow charts, systems and numbers,
Limitations, procedures V-speeds, and NUMBERS!
So it began in that ice cold room,
Cold as the grave, heavy with doom,
I watched the clock as my mouth rattled on,
I'm frozen in night, longing for dawn.
Soon I was walking, on to "the box!",
Time moved so slowly.... what's wrong with these clocks?
V1, Vr, "Fire in number two!"
Murphy you bastard! Now what do I do!
Then.... it was over, and I in a haze,
Emerged to the sunlight and one-eighty more days.
Till the next time, same place, different day,
"I have a few questions..." I'll hear him say.

–Richard L. Barlow

First Solo Flight

Yesterday I soared through the sky;
Saw the earth from an Eagle's eye.
Touched and warmed by rays from the sun;
My machine and myself transformed into one.
This brief time belonged to me...
Climbing higher, glancing back to see;
Where I'd been and What I'd done...
I realized how far I'd come.
Yesterday, a dream.Today, a reality;

My new awareness brought totality.
Today, I touched the sky. Today, I learn to fly!

–Betty M. Simpson

The Days That We Have Flown

You came to me,
like a frightened child,
and asked me how to fly.
I took your hand,
and led you up,
and showed you how and why.
We have looped and dove,
and rolled and spun,
and cut patterns from the sky.
And when we were done,
we'd had some fun,
and you learned how to fly.
Though you walk this Earth,
you've soared like an eagle,
seeing your shadow in the clouds.
You've done marvelous things,
since you've spread your wings,
and you've made your instructor proud.
And now that you have,
let go of my hand,
and set out on your own.
I hope that you,
will remember me,
and the days that we have flown.

–Timothy S. Bastian, 1997

Night Solo, Georgia

We practiced late into the night,
trying to keep the nose up
in the turn, back pressure,

back pressure to hold it
while the wings bent back around
through darkness and rolled out

touching pale blue taxi lights
a thousand feet below. Down there
the Georgia pines spread for miles

in all directions, down there
the Okefenokee Swamp sprawled
with all the snakes and gators

in the world. Yet there we were,
new as bait trolled on propellers
in a dark starless Georgia night,

trying to land our first time
solo, turning final,
gear-down and falling.

–Walt McDonald

Tilly

Her name was Tilly, she was one of the best
That night I put her to the test.
She looked so pretty, so shapely, so slim;
The night was dark, the lights were dim;

I was so excited, my heart missed a beat,
For I knew I was in for a special treat;
I 'd seen her stripped, I'd seen her bare,
I'd felt her over everywhere.
I got inside her, she screamed for joy,
That was the first night, boy, O, boy!
I got up quickly, as quick as I could,
I handled her gently, I knew she was good;
I rolled her over, then on her side,
Then on her back I also tried.
She was one big thrill, the best in the land
That B-29 of the Training Command.

–Author Unknown

Her First Solo

There was a student pilot,
her name was Jeannette.
She took off in a Tomahawk
and hasn't come down yet.

It was sometime in the morning
she began her solo flight;
but she's staying in the pattern
though the time is almost night.

She left in Five Nine Charlie;
her take off was quite fine.
She climbed across to downwind
and she kept that plane in line.

She does her pre-land checklist
and she makes her Easton call.

She throttles, trims, and uses flaps
and centers that there ball.

Then why is she still up there
going 'round and 'round the patch?
'Cuz she still can't see the glidepath
and that right there's the catch.

So sometimes it's a porpoise,
but most often a balloon;
and occasionally an outright high
as she comes in too soon.

Or sometimes it's a bounce,
or her airspeed's much to high
So every time she gives full pow'r
and jumps back in the sky.

The only thing I'm sure of is
that soon that plane will land
because thirty gallons soon runs out
in any pilot's hand.

She's going 'round another time,
but soon I think you'll see
the landing that she finally makes
will be "EMERGENCY!"

There was a student pilot,
her name it was Jeannette.
She took off in a Tomahawk
and hasn't come down yet!

–*Jeannette Nordquist Purviance*

Lesson Six Is Done

Oh, Master, let me fly with thee
And do thy will with accuracy
Keep my station on the tow
Straightened yaw string as I go.
Attitude first and attitude last
Eyes to airspeed briefly cast
Look outside and fix the nose
On the spot the Master shows.
Dive forty-five down to proper speed
Ease on four G's and let's proceed
Ease off pressure, keep it round
Pull G's again on going down.
To roll, it's aileron left to stop
Now forward, or the nose will drop
And all the way she goes around
And comes out facing the chosen mound.
Just pressure – never pull or jerk
Just pressure – it's easy gentle work
And use the rudder to keep the string
From sloppy, needless wandering.
And now for landing spot descent,
Hold it 'til the height is spent
Keep the speed; don't lose or gain
Hold it off; there, down again.
Now go and relax, we'll go again
And fly this lovely glider plane
And though ten lessons may be short
You'll have the basics of acro sport.

–*Tom Schollie*

Grounded

Because You Have Flown

John,
May the Lord give you
strength
to wing out in new
directions;
to seek other horizons
far beyond the reach
of a wingtip;
to feel the solid ground
at your feet
as did the first man
millenniums ago;
to walk through this life
knowing that you've accomplished
more than most men
Because you have flown...
And no matter the length of a flight,
all birds must come landing
back to earth...
eventually.

–Kathleen M. Rodgers, 1987 Alaska

Empty Tanks

The fog of old ships
Has enveloped his name,
The clear ice of time
Dulled the lift of the game,
The breaking of trail
In his ten thousand hours
Has emptied the tanks
Of his youth and his powers
And so he is through
And his license is gone,
The shadows of night
For the theme of his dawn.

He closes his log,
Takes the wings from his coat
With smoke in his eyes
And a lump in his throat,
But would he have quit
Had he foreseen the end
Or known from the first
He'd be whipped by the trend?
Not he! No! Not he!
May his breed never die!
His license may lapse
But his heart will still fly!

–Gill Robb Wilson, 1938

Religion/ Prayer

The Air Force Hymn

Lord, guard and guide the men who fly
Through the great spaces of the sky;
Be with them traversing the air
In darkening storms or sunshine fair
Thou who dost keep with tender might
The balanced birds in all their flight
Thou of the tempered winds be near
That, having thee, they know no fear
Control their minds with instinct fit
What time, adventuring, they quit
The firm security of land;
Grant steadfast eye and skillful hand
Aloft in solitudes of space,
Uphold them with Thy saving grace.
O God, protect the men who fly
Thru lonely ways beneath the sky.

–Unknown Author

This is the official United States Air Force poem that many cadets must commit to memory.

An Airman's Blessing

May the winds be calm,
And the weather fair,
And may you fly through the air,
with never a care.

For the clouds will float away.
And the heavens will smile,
As you lead your weary craft,
Down the last misty mile.

–*Misty Sorensen,* "The Mystical One"

An Airman's Hymn

When the last long flight is over
And the happy landings past
And my altimeter tells me
That the crack up's come at last,
I'll point her nose for the ceiling
And I'll give my crate the gun.
I'll open her up and let her zoom
For the airport of the sun.

Then the great God of flying men
Will look at me sort of slow
As I stow my plane in the hangar
On the field where flyers go.
Then I'll look upon His face,
The Almighty Flying Boss

Whose wingspread fills the horizon
From Orion to the Cross.

–Francis M. Miller

From the 18th Pursuit Group Songbook, 20 June 1940.

The Air Force Psalm

The Lord is my pilot. I shall not falter.
He sustaineth me as I span the heavens;
He leadeth me, steady, o'er the skyways.
He refresheth my soul.
For He showeth me the wonders of His firmament
For His Name's sake.

Yea, though I fly through treacherous storms and darkness
I shall fear no evil, for He is with me.
His Providence and Nearness they comfort me.

He openeth lovely vistas before me
In the presence of His Angels.
He filleth my heart with calm.
My trust in Him bringeth me peace.

Surely, His Goodness and Mercy
Shall accompany me each moment in the air,

And I shall dwell in His matchless heavens
forever.

–Author Unknown

Found in the historical archives poetry folder at the United States Air Force Museum, Wright-Patterson AFB, Ohio, with the Office of Chief of Air Force Chaplain's stationary and seal.

Seventh Heaven

I haven't done a lot of things,
But, harken, all you flying troupes
Do you suppose that Jesus Christ
Is into doing aerial loops?
I mean, you know He has the space
And perfect visibility;
I wonder if He has the time?
Well, just a whole eternity.

Has He felt exhilaration
At Mach 1 acceleration?
Heaven must have aviation,
Or I'll change my destination!

Does He rejoice up 'on the step'
For hours, and never falter?
Could He still fly without repose
And drink me under the altar?
Is He a stick and rudder God,
Or a delta-wing believer?
I hope I get a chance to ask,
'Cause I haven't done that either.

–P. A. Monahan

Military Service

FIGHTERS

Fighter Pilot

You can't come here, groundling,
I dwell in space so foreign
that even though you stare at it
you will never taste it.

While you are simpering over your greasy eggs,
I am climbing out at a hundred percent,
In burner,
Nose boring through the cold, blue-black air.
While you shave,
I make the sun rise and set again
with the touch of my gloved fist on the stick.

I can't see you down there
locked inexorably in twisting mosaic
beneath my wing.
You can't see me up here;
You don't tread among the gods.

This is a closed shop.
Only those who hack it are allowed,
and even those who dare had better press it;
For my purpose transcends aesthetics,
I'm here to flame something.

When I am on the ground reluctantly –
I seek the company of others who live
beyond the edge.

If I seem aloof and haughty,
Call it honest arrogance.

–Marshall Lefavor

For Don Morris: a T-28 pilot KIA, May 1970

Half a century gone. An image persists, born
When Spads and Fokkers dueled (they rarely fought):
Men with tiny airplanes, shined boots, smiles and scarves
(whine of wound-up engines, torque, smell and chatter of guns)
Watching for whims of weather or a blown jug.

Now, technology intrudes.
Pilots in their air-conditioned world seem far above all that.
Sometimes.

There are still a few who strap into small birds,
Call "Clear," cough from backfire smoke,
Groan off the ground and shudder skyward,
Bombs rigged with baling wire and wood blocks,
Hanging upon a prop.

These could care less about refueling tracks or radar plots.
Their need is to dodge thunderstorms, skim peaks, keep lead in sight.
Bomb-laden, to climb above the clouds is to stagger on the edge of a stall.
It is enough to reach the target, hit it, and return.

Their textbook written for an older war,
These pilots fight, and sometimes die, in this one.

It is good to know that some men
Still fly with scarves and laughter
As real pilots always have
And always will.

–John Clark Pratt, 1970

A Fighter Pilot's Friend

She has a language
All of her own,
Which you understand –
Now that you've flown.
She's made of something
More precious than gold,
And will always stay with you
If you don't get too bold.

Don't force her
Or push her,
She'll let you know,
As out on that mission you go,
When she's really ready
To give you her all;
And that's when you
Have to be on the ball.

Don't fear her – respect her;
And she'll treat you right,
And bring you home safely
From every flight.
Remember to thank her
Once in a while

For bringing you back
Over many a mile.

So love her, be kind to her
For there's nothing so great
As an ever – true friend –
A P-38!

–Author Unknown

The Flying Dutchman

Someone I flew with
has gone to the moon and back.
I haven't touched a rudder in ten years,
my only cockpit the family wagon
with a cracked windshield. Still,
I'm up there in famous make-believe,

but I'm dreaming of Stuart,
straight-in from 20,0009, Dave
who scooped a sand crane in the engine
but steered away from Waco
where he crashed, Keith
vaporized when a bomb
exploded in midair near Hanoi.

I might have died like Karnes
in Georgia, his first night solo.
He pitched too steep,
and the wings kept rolling.
When the controller called *Pull up*
Karnes jerked the stick back

tight to his groin, trying to obey
at two hundred knots, a thousand feet,
wings level by now, but inverted,
nose coming down, diving, someone
screaming *Pull up* in his headset.

Chandelles and loops, snap rolls,
tight Immelmanns
with a twist on top –
each turn was to get behind
somebody else, nothing beat
being second, a touch, a touch
on the button to bring him down.
The fun meant war.
The last aerobatic trick we were taught
was the victory roll.

Sometimes, squirting a thin jet
to water Texas grass trying to survive
another drought, I feel flames
in my fingertips and squeeze
to strafe begonias like a practice run.
At dawn, driving to work, I keep feeling
a tingling in my hands
and the weight on my backbone in a turn.

–*Walt McDonald*

To Live to Fly

On an airfield in Okinawa
a nine year old boy
with short blond hair
stood next to a fence

gaping at the sky.
An airplane had just taken off,
and from that day on
the boy lived to fly.
More than a dozen years later
the boy's dream came true
on the wings of the hot, dusty wind
over the streets of Larado.
Soon schoolboy ideals
were swept away
over the skies of Southeast Asia,
and the little boy was now
part-bird part-man,
a hawk-eyed loner
over the war land of the green.
He flew with a secret society
called RAVENS.
Their slogan "never more"
became the stuff
of legends and lore.
Today, a chunk of gold
circles his wrist;
it is a silent reminder
to a Raven,
that the most important thing for a man
is his freedom...
He has earned that privilege
and has served his country well,
since that day
the nine year old tasted the freedom
that the wind offers its flyers
when he sets his sights
to live to fly.

–Kathleen M. Rodgers, 1989 Louisiana

How Close Formation?

Wings nearly overlapped
We gently bank
Above the billowing clouds below.
Which stretch unending to the sight –
Their tops dark crowns to stormy threat
Now crouched in anger on the field.
Aware of this we circle in routine,
The patterns patient as we wait
Our turn to drop the brakes
And head toward earth.
When that time comes I must again
Resist the dawdling thoughts,
Force them from my mind
And concentrate on flying close.
Just tuck it in and count on lead.
He'll take me home,
He'll hold my hand.
Forget his acid tongue last night,
Forget the recent arguments:
The barroom bitterness of late,
The shoving for that coming job,
The push of wives who added fire
To points beyond their just concern.
Subdue the anger, mute dislike,
Pinpoint, instead, on getting down.
Blot out all thoughts that interfere,
Give him a signal it's okay –
That you again are side by side,
Again your confidence is high,
As one more time you two are forced
To link your fates
And drive yourselves
Into the furied elements.

–Don Clelland

The Searcher

The searcher
the seeker
the jack-in-the-bean-stalk dreamer,
clinging to schoolboy ways
and the adventure of Tom Sawyer days.
To taste the clouds
and ride the wind
in a whistling jet
is it such a sin?
To be so young. . .
and yet a man.
To pierce through the blue
in search of that dream;
to yearn for the best
on the tip of a wing.
The young fighter pilot
ask only to fly
to seek all the wisdom
up in the sky.
To hunger for a dogfight
to thirst after dreams.
To be the searcher
the seeker
The dreamer of things...

–Kathleen M. Rodgers, 1987 Alaska

Last of the Breed . . . "An Old Head"

He's the last of the breed. . .
"An old Head"
He's been around awhile
among that brotherhood of flyers.
He's a teacher – a Boy Scout leader
to the young cubs
fresh out of pilot school.
He's a permanent fixture
at the O'Club Friday nights.
He's a crusty old bastard
who's earned the rite of passage
in the hell and glory of combat.
He's not some GI Joe
cranked out in a factory
to be tossed away – forgotten
when his flying days are through.
He's the old Spartan warrior
never giving up – or in.
For his flyer's soul still soars
beyond the wounds
and retirement.
He's a legend
in the fighter community,
the best there was. . . and is.
The Last of the Breed. . .
"An Old Head"

–Kathleen M. Rodgers, 1987 Alaska

When the Mustangs Came

In the darkest days of W.W.II, the Luftwaffe roamed at will
Across the skies of fortress Europe, to terrorize and kill
And far below them thousands died. As tools of war were hurled
By those who swore allegiance to, a thug who'd rule the World
Death was not selective, the innocent and helpless died
The young who'd never lived,
and moms with toddlers at their sides
In the air, on the ground, aboard the ships at sea
Those who served the tyrant sought, oppression of the free
But factory's where their guns and tanks,
their planes and ships were made
Would feel the awesome, righteous wrath, of allied bombing raids
But as the bombers struggled through,
without escorts at their side
Some raids would be successful, and some were suicide
With targets beyond the range, of thunderbolts and '38's
The bombers they were easy prey, as the Luftwaffe lay in wait
But that all changed when the Mustangs came,
a drop tank on each wing
The fuel and range to carry out, the vengeance they would bring
As American sons in '51's, prowled cold and hostile skies
The tables turned, the Luftwaffe learned,
their role was "compromised"
With Mustangs flying overhead, mile after bloody mile
Roles reversed, no longer first, the Luftwaffe was on trial
Bomber crews in crippled ships, were no longer left alone
Like guardian angels the '51's, were there to nurse them home
And as they faced the Luftwaffe guns,
and sheltered bomber crews
With valor worthy of their cause, the Mustangs paid their dues
Some returned without a scratch, some would crash and burn
Others barely made it home, and some would not return
On combat missions in a realm, that few can understand
With no comrades there to comfort them,
or hold their dying hand

At thirty thousand feet, and so many miles from home
It was the fighter pilots fate, to fight and die alone
And half a World away, to the land of the rising Sun
The allies of the Nazi cause, would face the Mustangs guns
B-29's enroute to bomb, the warlords of Japan
Could do their job and make it home,
with the Mustangs close at hand
Then mushroom clouds above Japan, and W.W.II was done
Ignited by the Lords of war, but won by freedom's sons
Then 1950 Korean skies, were war torn and aflame
And like before the knights of war, the gallant Mustangs came
Each time the free world was at risk, in wars around the world
Mustangs always fought to keep, free nations flags unfurled
And in the hearts of grateful men, there is a surge of pride
As they recall the gallant ways, the Mustangs fought and died.

–Ivan L. Fail, 1986

March 10, 1966

In A Shau Valley
that cloudy day,
a downed pilot
began to pray
with many Viet Cong
all around
and from Air Rescue
not any sound.
His doom was coming,
he had no doubt–
when out of a cloud
came one lone scout

who hazarded one pass
and landed then
on a scarred runway
scattered with tin.
The bullets did whiz
both right and left
while he accomplished
an august theft–
and was back in the air
at minimum speed:
not too many
are of that breed.
And both pilots know
they shunned the sod
not by luck, but by
the grace of God.

–James A. Grimshaw

Phantom

Yesterday I raced a fleeting shadow,
swift as life
It hurtled over sunlit fields and lush
brown hills;
With careless ease it leaped the rivers,
highlands, woodlands, and the
scattered towns.
A thing of beauty, sun-born, wild with
speed
It foiled the clutching fingers of the
grey mesquite.

I landed – and the Phantom came to rest
beneath my wings.
I felt that I had killed a thing of life
and mourned its passing.

–Major Paul

Sweeping Squadrons

Filled the summer sky
white trails across the
brilliant blue
We met them head on
five miles high
They were many, we were few
Went the day well?
We died and never knew
but well or ill – freedom
we died for you
And left the vivid air
signed with our honour
And now –
Do you remember us
they called The Few?
We need to know that we
are not alone
That here and now our
sacrifice is known
And we are not forgotten

–Unknown Author

An inscription found by the graveside of an unknown pilot – Killed in Action.

What Is a Fighter Pilot?

A fighter jock is quite a phenomenon. He likes flying (single seats only) and especially gunnery, acrobatics and cross countries. He has a strange fascination for flying boots, gambling, cigars (the bigger the better), and breaking glasses. He can usually be found in sports cars, at parties or happy hour. His natural habitat (while on the ground) is the Land of the Bearded Clam, Europe and/or certain parts of the Orient. He has an affinity for women and booze (especially Martinis so dry the bartender just faces Italy and salutes). He likes Steve Canyon, to read Snoopy, eat steaks and tell dirty jokes. His favorite hiding place is in dark cool bars or behind a pair of dark glasses. He is capricious. To amuse himself he may fire practice flares from mobile control, throw empty beer cans down the BOQ corridors, pour drinks down an over-exposing décolleté, or become generally obnoxious. His favorite conversation revolves about a continuous chatter concerning flying, booze or females (the order of priority is apparently irrelevant).

He has an aversion for survival training, bomber pilots (or most other pilots for that matter), mobile control, AO duty, and extended alerts. He tolerates ankle biters and house apes (other than his own) and has an overwhelming hatred for bingo. Whenever possible he avoids weather, icy runways, lost communications, flame outs and ejections. Water makes him sick (unless frozen and surrounded by Scotch), and would rather face a firing squad than be caught pushing a baby buggy or carrying an umbrella. At the mention of matrimony, he becomes a catatonic schizophrenic and has a mysterious distaste toward wearing a wedding band.

A fighter pilot is a composite. He has the nerves of a robot, the audacity of Dennis the Menace, the lungs of a platoon sergeant, the vitality of an atomic bomb, the imagination of a science fiction writer, glib as a diplomat, impervious to suggestion and is a paragon of wisdom with a wealth of unassorted, completely unrelated and irrelevant facts. He wears the biggest watch, has the shortest staying power and is always trying to get laid on credit. When he tries to make an impression, either his brain turns to mud or he becomes a savage, sadistic jungle creature bent on destroying the world and himself with it.

Who else can cram into one flying suit: check lists, maps, Zeus openers, check lists, a dime novel, knives guns flares and snares, nylon cording, a handkerchief, assorted inhalers, aspirin, cigarettes, a flashlight, check lists, pencils, pens, gloves, a deck of cards, coded telephone numbers, a wallet, keys, his horoscope, a talisman, a St. Christopher medallion, check lists – and a chunk of unknown substance.

At home with his wife he is docile, sweet, tender, loving, amiable – just a helluva nice guy to have around the house – straight arrow all the way, except when they're fighting – then he becomes a beast who is tyrannical, suspicious, diabolical, and a masochistic sex fiend who just ain't got no couth (these symptoms may also appear after beer call).

As a father he is tough but oh so gentle, kind, just, protective, far sighted, ambitious and really proud of that young fighter pilot (he'll never admit it, and it's never displayed in public, but that goes for the little girl too).

In the air he is calculating and confident. His voice gruff and steely cool (an acquired characteristic regardless of how he feels), pierces the garbled waves, baking terse commands. On the hunt he becomes part monster: scanning with the eyes of a falcon, has the reactions of a cat, the instincts of a barracuda, the cunning of a fox – and the ability to rotate his head 360 degrees on all axes. When approaching the target, mind and metal fuse, spawning a killer-child. Destruction is sure and precise as Euclidean geometry. Steel and fire split the icy atmosphere – swift and merciless he revels in his private moment of truth.

After the mission he is tired, thirsty, dirty and bedraggled. He walks with his legs crossed to the nearest latrine (or empties out his G-suit). Hair matted with helmet rat snarls and mask scars etched on a red, raw face, he knows he has bid and beaten the grim reaper. And then with the oily odor of JP-4 clinging to a salt encrusted zipper-ripper, he'll unleash that shiny-eyed smile which says "let's press on to the O club and inhale a few tall frosty ones" – whereupon he miraculously regenerates into a critical mass and with flurry of hands, arms, legs and body english stuns his alcoholic cohorts with tales of "hairy" deeds.

A fighter jock is magic, a master impostor, Houdini with the top of his blouse unbuttoned. Sometimes he's old, sometimes young. Immature yet sage. He is instant fear and lasting bravery. The original metamorphosis. Hovers between play and business, and can make your date vanish right before your eyes. He is present, past and future rolled into one. But most of all he's got wings – with a throttle in his left hand and the stick in his right – shackled to a million dollar blow torch and always ready to get the maximum out of every minute of every hour of every day.

–Ford H. Smart

The Fighter Pilot – a Tribute

Say what you will about him–arrogant, cocky, boisterous, and a fun-loving fool to boot. He has earned his place in the sun. Across the span of fifty years he has given this country some of its proudest moments and most cherished military traditions. But fame is short lived and little the world remembers.

Almost forgotten are the 1400 fighter pilots who stood alone against the might of Hitler's Germany during the dark summer of 1940, and in the words of Sir Winston Churchill, gave England, Its finest hour. Gone from the hand stands of Duxford are the 51s with their checkerboard noses (78th Fighter Group) that terrorized the finest fighter squadrons the Luftwaffe had.

Dimly remembered–the 4th Fighter Group that gave Americans some of their few proud moments in the skies over

Korea (and over Europe in WW II). And many a fighter jock from WW II had to do it all over again in Korea. It is almost a forgotten war; the one between World War II and Vietnam. That was when the United States still fought aggression. But it also marked the beginning of no-win policies that would dramatically weaken the free world.

How fresh in recall are the Air Commandos who valiantly struck the VC with their aging Skyraiders in the rain and blood-soaked valley called A Shau. And how long will be remembered the Phantoms and the Thuds over Route Pack Six and the flak-filled skies above Hanoi?

Barrel roll, steel tiger, and tally ho; so here's a Nickel On The Grass to you, my friend, for your spirit, enthusiasm, sacrifice, and courage – but most of all to your friendship. Yours is a dying breed, and when you are gone, the world will be a lesser place.

–Raymond B. Tucker, Lt Col USAF

BOMBERS

Ode to a Bombardier

On a lonely road, through a cold, bleak night,
A grizzled old man trudged into sight;
And the people all whispered over their beers,
There goes the last of the Bombardiers.

What's a Bombardier? There came no reply,
The men turned silent – the women sighed;
As death-like silence filled the place,
With the gaunt, gray ghost, of a long, lost race.

It's hard to explain, that catch of breath,
As they seemed to sense the approach of death;
Furtive glances – from ceiling to floor,
'Til something, or someone opened the door.

The bravest of hearts turned cold with fear,
The thing in the door was a Bombardier;
His hands were bony, his hair white and thin,
His back was curved, like and old bent pin.

His eyes were two empty rings of black,
And he vaguely murmured, "Shack, Shack, Shack."
This ancient relic of the Second World War,
Crept 'cross the room and slouched to the bar.

They spoke not a word, but watched in the glass,
As the broken old man showed a worn bombsight pass;
And hollow tones from his shrunken chest,
Demanded a drink and only the best.

The glass to his lips, they heard him say,
"The bomb bay is open–the bombs are away;"
With no other word, he slipped through the door,
– And the last Bombardier was seen no more.

–Author Unknown

From *The War Chronicle of John Hank Henry*, a B-25 pilot in World War II.

Today I Rode A Legend

Today I rode a legend, and sat where heroes died
Tho humbled at just the thought, still I felt a surge of pride
As those sweetly singing cyclones,
thundered swiftly through the sky
I was carried to the past, to an era that's long passed by
Yes the pages of the past, they seemed to live anew
As in thought I paused in reverence,
to these gallant planes and crews
From somewhere a ghostly presence, arose to prod my thoughts
To remind me of the freedom, that blood and valor bought
And as my memory wandered, and haunted me this day
I thought about the debt, I can never hope to pay
I thought of all those men, faceless and unknown
I pondered at the fate of those, who would never make it home
And I thought of gallant crews,
who nursed back their crippled ships
And anguished for the others, who went on those one way trips
How many were only kids, still fuzzy cheeked and green?
Who would die before they'd live, in their flaming Seventeen's
Others were hardened Vets, although still young in years
Men tempered in the fires, of danger, death and fear
Men who left behind, loved ones to mourn their loss
The evil make the debt, the innocent pay the cost
Uncommon valor a common virtue,
among these planes and men
They fought and died, so such as I, could live in peace again
Yes today I rode a legend, to whom I owe an awesome debt
That we must never take for granted, nor in complacency forget
Awards for valor were commonplace, in this aging angel's realm
For all her gallant crewmen, and the pilots at her helm
As she fought to guard our freedom, she earned with awful loss
The Medals of Honor –, Purple Hearts –, Stars of Gold –,
And Distinguished Flying Cross
And telegrams of condolence, to the loved ones back at home

Of the gallant lads who fought and died, in a foreign hostile zone
Scorned as a "Tool of War" by some,
who've not heard an angered shot
Yet they enjoy the freedom, that her blood and valor bought
Yes she's a warrior true, but deserving of respect
Not the scorn of those ungrateful, their hatred, nor our neglect
She's American through and through,
and she did her mission well
Far above the call of duty, in her missions over Hell
Yes today I rode a legend, a lady of whom I'm proud
Who in the face of death and danger,
never cowered and never bowed
A veteran worthy of allegiance, the love of all the free
A gallant bird who fought and bled, and died for you and me

–Ivan L. Fail, 1986

Anamnesis of a Bombardier: My Father, Sleepwandering From the War

The veteran Greeks came home
Sleepwandering from the war...
–Edwin Muir
"The Return of the Greeks"

Falling back upon the saturnine year,
arresting memory's absolution
'midst sleepy breezes near century's end,
come remembered strides on a bomb bay spar
unfettering an iron-cased judgment.

Then came the day your name failed the roster,
You – bought by Fates for unreturning men
who, in your absence, sinned in irony –
colliding with an enemy fighter.
Men still falling…upon your Stateside years.

–Major Jeffrey C. Alfier, 2001

Tribute to the Queen

From Guadalcanal and the Philippines,
at the start of W.W.II,
To the hostile skies of Europe,
thru miles of flak she flew.
At home at thirty thousand, majestic as a Queen,
A Silver Bird flown by men,
many in their teens.
She carried war to the tyrant's lair,
to keep all nations free;
She flew thru flak and flame, as far as the eye could see.
She slugged it out with Hitler's best,
brought her dead and wounded home.
Damaged and with engines out, it was often times alone.
Born of war, but seeking peace, she carried valiant men
into the very jaws of death, and brought them home again.
Berlin, Frankfurt and countless others, courageous daylight raids,
and only God in Heaven knows, the awesome price she paid.
She met death at 30,000 or on a treetop run.
A victim of ack-ack shell, or Luftwaffe fighters gun.
Like all the men who flew her, for peace and hope she yearned,
but too often mission boards would read, "Failure to Return".
Often plane and crew went down, in a hostile place.
Others were missing in action, and lost without a trace.
Her era's in the past, but the history that she made,

must always be remembered, and never be betrayed.
Generations have come and gone;
Enjoyed their hopes and dreams,
Yet never paused ingratitude, to this aging Silver Queen.
And the men who flew her, Heroes everyone,
who stood between our nations shores,
and the tyrant's mighty guns.
Yes, she's tired and weary, A little aged and worn.
But, she fought and bought their freedom,
before most of them were born.
And we who still remember, To-Jo and Hitler's dreams
Stand a little prouder. . .
in the presence of the Queen.

–Ivan L. Fail, 1985

The Liberator

All out war was raging, the flags of hate unfurled
A knife was held at freedom's throat by the tyrants of the World
In Europe millions perished neath the Fascist-Nazi tide
And throughout the vast Pacific, countless others bled and died
Pearl Harbor lay in ruins, with hundreds dead and maimed
Into this carnage and despair the "Liberator" came
Four Pratt and Whitneys on her wings would thrust her
through the sky
And take her to the gates of Hell where she'd fight and often die
Freedom's never "free" and with her gallant crews
She would fight to keep us free and she'd pay some awful dues
She tracked Pacific targets, the warlord Tojo's lair
She braved the ack ack from the ground and the zeros in the air
From Guadalcanal to Okinawa and the endless hell between
From small volcanic atolls to jungle islands thick and green
To the bloody skies of Europe

where the Luftwaffe reigned supreme
Men who'd die a futile death to fulfill a man's dream
She bombed the tyrant's lair, hostile and remote
And often times her gallant crews were "kids" too young to vote
Berlin, Schweinfurt, Frankfurt, God knows how much more
Who understands the madness and the fickle winds of war?
But like the noble comrades she fought and died among
Her role is oft neglected, forgotten and unsung
Another date with destiny these gallant warriors kept
Ploesti – in Romania where the angels must have wept
Each mission took its toll, in planes and ravaged men
But until her job was finished, she'd return to fight again
And far below within his lair the tyrant cringed in dread
As formations of B-24's would thunder overhead
In shot up planes, wounded men, some overdue and lost
The years have come and gone, the World is not the same
But grateful nations still recall when the "Liberator" came
She wrote her page in history with the blood of valiant men
And God forbid that such as this should ever be again
It seems I hear her thunder like echoes from a ghost
Through the vapor of the ages I still see her at her "post".

–Ivan L. Fail, 1988

Gaining Altitude

Was it your hate of war that lay behind
the silence of thirty combat missions?
Grown fatalistic, you scorned the order
not to remain in the nose for take-off.
In flight, you cued off the lead bombardier
and called it war. Or maybe you just blamed
the genius of intervalometers,

picked up your lamp, and followed the phantoms
which did not speak until the pins were pulled...

Spring. Your wit grows acrid. I forgive you.

–Major Jeffery C. Alfier

WAR

And I Look Down

A magnificent view, I tell you!
The engines of the aeroplane drone complacently,
and I look down
with hungry eyes
at the land below.
I see rice paddies, delicate,
like stained glass windows of tasteful
greens, browns, blues,
a tactful blending of man's handiwork
with nature's,
But pockmarked by one of man's diseases,
war.

I see mountains, gnarled,
Like the hands of an old Asian farmer,
rough, rugged, taut,
With long, lean, fingers reaching out
towards me,
Some of the fingers burned
by the touch of war.

I see beaches, dazzling,
Narrow buffer zones attacked by both
land and sea.
Long, lonely beaches and virgin dunes
are teasing...
Their nude forms beckon me to
take a break from war.

And even the sky, which is
Such a soft sea in these climes...
above, ahead, below,
Complimenting the eye with moody color...
is often
Invaded by black, billowing smoke, a flag of the war.

A magnificent view, I tell you!
The engines of the aeroplane drone
complacently,
And I look down
with hungry eyes
at the land below.

–Roland Herwig

The Return

Five minutes from target,
The auto-pilot set;
The peaceful drone of engines,
Hands clammy, forehead wet.

Our attack was very successful,
We caught them by surprise;

We bombed and heavily strafed them,
Smoke poured up in the sky.

We may have left destruction,
In planes and bodies, too;
But here beyond the target;
There's the quiet of the blue.

The sky's in all it's glory,
Of floating, fluffy clouds;
The large one rears it's cotton head,
The smaller ones it crowds.

And just beneath the water,
A melancholy blue;
It's waves are small, it's flashes bright,
As sunlight filters through.

We're at war, we know it well,
The targets just behind
But flying now, in nature's realm;
Her restfulness is kind.

For nature cares not if there's war,
The world is hers to roam;
And now our mission is complete;
It's peaceful going home.

–Author Unknown

This poem was written by a copilot enroute to his airbase after a minimum altitude strike against Mapanget in the Celebes Islands on 21 November 1944. Unfortunately, no names were provided. It first appeared in *The War Chronicle of John Hank Henry,* September 1940 –October 1945.

When It's Over

When the cannon's last roar echoes on the hill.
When the last bullet has found its target.
When the last grave has its fill,
And the last orders are in the pocket.
What will be left?
When the last bomber and fighter take to their wings.
When the last club no longer sings
With the fears of men and the lusts of women
And the runways are covered in dust.
What will be left?
When the last man dies in fear.
When the last woman sheds her tear.
When the last fool spits on a flag.
When the last profiteer pockets his swag.
What will be left?
Hate,
For wars only reinforce it in this world.
Poor gain for such sacrifices unfurled.
And before these scars heal,
At Mars' altar our sons will kneel.

–James H. Tiller III

*The Wingmen from Takhli**

When time passes on, and we have
reached the twilight of our lives,
I shall harken back to over a hundred flights
in these war-torn Asian skies and,
Once again, I'll hear that roar of burner,
blast of cannon and screech of tires.
Through weak and misty eyes in vain will I look

and search the skies
For those wingmen no longer here:
And always, my heart, my soul and my memory
will take me back...
Perhaps to Quang Khe, or maybe Dong Hoi† . . .
but always
To those wingmen from Takhli.

–Eugene Cirillo

* A base in Thailand. † North Vietnamese bases.

Precision Guided Munitions – Fourteen Second Flight to Dispossession

Climaxing the solicitude of near-term victory
we spawn fatherless, children of the mourning ashes
to beg the phantasms of a killing distance.
Unconscionable symbiosis of our
children's games and airmen's laser fuzes:
the faceless name of terminal guidance.
Yet we think that God is impressed
by the heartless protractions
we set in the heavens
'twixt the breathless quick
and unwhite dead,
in the wars
without
tears.

–Major Jeffrey C. Alfier, 2000

War Stories

The years pass so swiftly,
How quickly they fly!
As if pushed by the jet stream
In yonder blue sky.
And lately, it seems
I'm found more and more,
Retreating to memories
Of past glory and gore.

Which finds me accompanied
By aviators rare,
Their eyes wrinkled and browned
From the sun's naked glare.
And I listen to stories
Of War I and War II,
ICD, Korea, DC-1, DC-2.

I am constantly speechless,
Enraptured, in awe
Of War Stories witnessed
That I never saw!
And I wonder, in silence,
These stories of skies:
Which are the truth,
And which are White Lies?

So I developed a system
To sort wheat from chaff,
When to cry silent tears,
And when to just laugh.
You listen, impassioned,
To each sacred word
Of skies full of fire,
Of wounded, sick birds.

But you watch him intently,
A scotch in his hand;
His hat cocked so smartly,
His smile smooth as sand,
One hand on his hip,
(or somebody's thigh):
Watch his lips; if they move,
Then you'll know it's a lie!

–Michael J. Larkin

Soaring

The Voice

The sky before me is full of power. Cumulous clouds stretch out ahead for hundreds of miles casting an abstract patchwork of shadow over the hot desert landscape below. They begin their brief lives as swirling shreds of white mist and grow into bulging, spastic shapes that rise up into the clear, cobalt blue sky. Some fade away, while others gather together and become one. Upon meeting, some grow like a cancer into frightening things to become nature's demons. These convulsing masses spit out shards of spent water vapor in all directions heaving violent wind, rain and electricity. But some clouds billow with beauty and paint an intoxicating picture, which lures me beneath. I can feel the energy of the day and move towards the power that will embrace my graceful white wings.

I am alone, in my sailplane flying fast through this maze of beauty and destruction, silently cutting a path beneath these unruly clouds. Huge columns of invisible air rise and fall violently all around me. I can feel, and hear, the friction of the wind flowing over the sailplane's polished skin. The energy outside my cockpit presses against the control stick as if it were a living being and my hand responds with deliberate, but tenuous moves.

I feel a shuddering rumble as the sailplane slams into a wall of rising air that's being sucked up into the base of a darkened cloud above. Wings flex wildly and I'm pressed into my seat as I pull back on the control stick. The sailplane climbs higher and higher, one thousand feet up until its skyward momentum runs out. Now the cockpit is filled with an eerie silence, as the powerless craft can climb no more. Pushing the control stick forward, I dive for speed, which once again starts the vital flow of wind racing over the fiberglass skin.

My body seems to melt into the fuselage, pushing out through the physical boundaries of the sailplane and into the magical world outside. The mercurial space before me is a mysterious expanse of torrent and calm, but somehow I know where to go. Because it is in this place, far above all that is familiar, that I listen to my soul and feel a true sense of my own spirit. I am alone in this sky of dreams, feeling the power of God. Here, I follow no one, and here I listen only to myself. It is at this moment when I first hear the voice from deep within me crying out; "Here I am and I am with you!"

This was that day when I first met my true self.

–Chris Woods

Goal

Hip hip hooray, it happened today
I finally made it to goal
After days bombing out, plagued by self doubt
It's certainly good for the soul
After releasing from tow, I got really low
And was staring at a bomb out score
Circle in zero sink, take some time just to think
Then finally jag a good core
Up and away, this could be my day
So I race along on down the line
Tonight I'll get pissed, but right now I need lift
Well everything's going just fine
But halfway about, my luck just runs out
Once more I am getting real low
So in zero sink, I stop and I think

Damn it I'm going too slow
It seems that my plan, of being top man
Has just gone right out the back door
With a smile on my face, I accept with good grace
The fact I won't win anymore
Because flying XC, is still fun for me
Though many would think me quite mad
To lose or to win, is not really the thing
And if you don't get it that's sad
Well over the line, doing a hundred and nine
Kilometers I'm talking not miles
Pull up from the ground, and crank it around
Into goal for a beer and a smile.

–Dr. James Freeman

Gems of Wisdom

My friends, these words I now relate
True gems of wisdom carry.
Though words would ne'er a flier make,
I'll share with you this savvy.

Discard the cu that's done its due.
Don't further dilly-dally.
A cu that's young and wide awake,...
Now that's the place to tarry.

Nor steer a course that's straight and true.
But go where lift is carry.
To try to cross a hole that's blue
Could be your grand finale.

Now when these things you truly do,
And when you ne'er miscarry.
Please share with me your secrets too,
And I will forthwith sally.

–Jack Greene

Cu stands for cumulus clouds, which are formed by rising warm air. Pilots search for cu's as potential sources of lift to sustain their flight.

Sky Fever

I must go up to the Owen's again, to the lonely mountain sky,
And all I ask is a strong wing with strong arms to steer her by.
And the rotor's kick and wind's song and the tight sail shaking,
And the gray cloud mist on Mt. Whitney's face
and a rosy dawn breaking.

I must go up to the sky again, for the call of the mountainside
Is a wild call and a clear call that may not be denied;
And all I ask is a boomin' day with the big clouds flyin',
And a thermal day with the "dusts" away, and the eagles cryin'.

I must go up to the sky again, to the vagrant pilot's life,
To the mountain's way and the eagle's way,
where the wind's a whetted knife;
And all I ask is a merry yarn from a laughing fellow rover,
And a quiet sleep and a sweet dream when the long flight's over.

–Karl Stice

Soaring Soliloquy

Life is but a beach of sand,
some billion grains of play.
An endless, magic wealth of choice.
How swift life sifts away.

So search, so seek, so reach for stars.
Fear not you'd go astray.
Life's spoils will fall to those who quest,
ne'er those who shun the fray.

So cast aside the bounds of earth.
Then fast be on your way.
Some far off place might be your goal.
Some lofty peak your play.

And when the sands of time run thin,
those final grains might say.
Please God, he begs one final task
to close an earthly stay.

Then clouds will serve as silent stairs.
Soft winds will show the way.
To reach the stars, to touch the sky,
to close that final day.

–Jack Greene

A Place To Play

Somewhere, there's a place to fly
Where winds will bend towards the sky.
Where close at hand, the ridge will say,
Come follow and we'll play.

Somewhere, there's a place to fly
Where clouds will form, and build, and die.
With lifespan short, they show the way
That you and I might play.

Somewhere, there's a place to fly
Where silky waves will lace the sky.
Transfixed like stunning crowns, they say,
Come hither and we'll play.

Somewhere in that endless sky,
There're places low, amidst, or high
Where nature's forces e'er hold away,
And show us where to play.

–Jack Greene

Dreams

Dreams are filled with towering clouds
That burst across the sky.
Great balconies of lofty lift,
A million things to try.
Where mystic, magic, silky waves
Ascend in awesome might.
With endless streets and snow-capped peaks
And winds that never die.

Dreams are filled with wondrous flights
That journey far away.
That just go on, and on and on,
And on 'til dusk of day.
To twist, to turn, to spiral there,
perchance to touch the sky.
Or much too soon to tumble back
From where wild dreams would stray.

Dreams come and go. Dreams just pretend.
Yet dreams e're grace the sky.
Such simple funful fantasies,
Those things we deign to try
Yet in the end, one theme transcends
All mortal soaring thoughts.
When God unveiled this place called earth,
He meant that man would fly.

–*Jack Greene*

Parallels

Into each day some darkness is cast.
All around us, so cold, so dreary.
The burdens of life seem endlessly vast.
Just being so endlessly weary.

Into each flight some struggle will fall.
The sky is so still, so lifeless.
To move e'en at all would be at a crawl.
Each nibble of lift is priceless.

Into each life must courage be born.
So struggle, so stay with the race.

To fall by the side, cast aside with scorn.
When it's done, but yourself you must face.

–*Jack Greene*

It's Easy to Forget

Through pathless skies you sweat and strain
The countryside you roam.
The final turn point's now at hand,
That joyful turn for home.

So many miles have been traversed.
What lies ahead ... unknown.
Let patience be your firm resolve.
The task must still be flown.

The day is late. The sky is blue.
The air so smooth, so still.
A knot of lift is all you beg.
'Tis luck you need . . . not skill.

What was that? You sense a change.
You feel a gentle bump.
Could this be it? Which way to turn?
The needle takes a jump.

From depths of despair . . . euphoric delight.
You slowly gain some height.
Those agonizing last few turns.
The finish line's in sight.

The cruise for home is joyful bliss.
You're smug about this flight.

Right from the start, you had it made.
Of course, you planned it right.

–Jack Greene

The Glider Pilot's Lament

Ask me no questions, and I'll tell you no lies
I bombed out today, fell out of the sky
I was heading down course, as fast as I could
With everything going, just as it should
Into a thermal, hear the vario scream
Crank it on over, you know what I mean
I'm heading for heaven, at better than seven
Then pulling the pin, as I pass through eleven
Out of that thermal, out onto glide
Feeling at ease taking all in my stride
Straight down the course, at best speed-to-fly
Sink alarm winging, I start to ask why?
I haven't had shit, for over 10 K
And I'm sinking out fast, I see with dismay
A thermal, a thermal, a thermal I need
I put on the brakes, back off on the speed
Then I hit a bump, relax or you'll lose it
I crank it around, trying to use it
Still sinking out, it seems so unfair
Pick out a landing, just over there
An eagle, and eagle, an eagle I'm saved
But as I watch, it's just not my day
This eagle it seems, wasn't going to goal
No great surprise, that isn't his role
He circled on down, to land in a tree
Quite closely followed by glider and me
So if you ask me, "What happened today?"

This is 'bout all that you'll get me to say
Sometimes you win, and sometimes you lose
To climb or to glide? You get to choose
So I didn't win, although I planned it
I bombed out today, and f**cken well landed

–Dr. James Freeman

My Thermal

'Mid pathless skies, I groan and grieve.
I ponder, should I smartly leave
My thermal.
The earth below, such harsh despair.
What twist fate would drop me there?
My thermal.
To stay? To go? I vacillate.
I find myself in sinking state.
My thermal?
I pout and spout and cast about.
It's clear now, I am quite without
My thermal.
What can I do. When will I learn
Just when to stay or when to spurn
My thermal.
I'm now at rest in ugly weed.
Indeed because I failed to heed
My thermal.

–Jack Greene

Absurd Possibilities

For reasons that abound,
I'm always first one down.
Today I'll strive to be,
The last one on the ground.
But downed I'll take with glee,
Compared with what could be.
Like last frog in a pond.
Or last fish in the sea.
Thank god my Maker found,
To not turn things around.
Suppose He'd have me be,
The last hound in the pound.

–Jack Greene

Will I Ever Know

One day, I glanced toward the sky.
A far off flash that caught my eye
Would twist and turn and frolic there.
Then drift and drift and drift . . . but why?
What purpose for such aimless play.
Perchance to find a place to stay
And climb a magic stairway there.
Caress each cloud . . . then race away.
Whence from he came? Where would he go?
What kind of game with pace so slow
By chance would have him pass my way?
These things . . . are these for me to know?
And when he paused straight overhead,
That slender craft with outspread

Would seem to have a purpose there.
He spiraled up . . . then off he sped.
Enthralled, I watched his funful play.
Yet tears my sadness would betray,
For soon, a final flash of white
Would tell me . . . he was gone away.
I wondered what his thoughts embraced.
The lonely flight? The soundless grace?
Untouched in tranquil solitude.
Oh someday . . . might I share his space.

–Jack Greene

The Sad Sack

One day I thought I'd reach the wave,
My high point was the notch.
To fly away, now that I crave,
Before I leave, I'm lost.
To take a tow, I must be brave,
My ropes all frail from rot.
I never try heroic saves,
My saves are always flops.
One day I proudly stood on stage,
The part I played ... a prop.
I never ever risk a shave,
My throat I'd surely chop.
If I should someday be a slave,
I'd slave for slaves for naught.
Or should I hide inside a cave,
That cave would on me drop.
Now fellow friends, I don't make waves,
I know my humble spot.

But once before I'm laid to grave,
Might once I find the top.

—Jack Greene

The Myriad Scenes of Soaring

If all you ask is just a chance, to soar one summer's day.
To twist and turn, to drift awhile, yet barely stray away.
To climb, to chase a puffy cu, to simply stay aloft.
My friend, you've found your special thrill.
I like your choice of play.
But rather, would you speed away to race across the sky.
Where for the moment, friends are foes and with each other vie.
Where every move is borne of speed,
to dash from here to there.
Then e'en more swiftly run for home.
These things are yours to try.
Perhaps the wild and thrashing ridge,
would be the place you'd be.
With every sense of keen alert, you skim a million trees.
Where speed and furry mark the way. Where violent forces play.
You rarely stop to take a turn. Is that the place you'd be?
Still others choose to sail away, beyond where mortals play.
With ridge and wave and thermals there
and winds to point the way.
The thrill is just to move along as far as eye can see.
Sail on my friend. Sail on and on 'til twilights close your day.
Then some would seek a silky wave to touch a topless sky.
Where shimmering stillness sings its song,
yet dangers lurk close by.
With lennies stacked in awesome might, where lonely vigils keep.
You're fixed in mystic magic trance, this place you choose to fly.
The sun dips low as evening falls. The day has done its due.

And one by one, we're coming home to tell of tasks we flew.
We feel a special kinship now. We share a common bond.
It matters not your task that day, but was that task for you.

–Jack Greene

Soaring Verse

Flight ... a fascination of man
For me it has been a lifetime past time
But, my flying experience is slightly different
than most envision.

For the sport in which I take part in, is known as soaring.
Flight . . . without the aid of an engine
People ask is that at all possible?
My answer . . . the birds can do it!
For essentially that is what we become.

Even though in this form of flight
we are on a never-ending search for lift
in order to remain in this new world.
However, work is not
For at the same time the peace of flight is soothing.
For in soaring, the person becomes part of the plane and thereby
leaving their earthbound problems behind.

Now is the time to relax, let time fly by,
Listen to the wind whistle by and
take in the view that the birds see daily.
For eventually your time is over
and it is time to return to the Earth below.

–D. Elber, 1988

For Children

To a Pilot's Son

Though my job takes me to far-away places,
Far from home among many new faces,
Enjoying rich, glowing sunsets and brilliant sky of blue,
The sad part is it takes me away from you.
But to a pilot a plane is a mixed blessing,
Like cold, sleepless nights and 4 o'clock dressing,
All the nights missed tucking you in bed,
Too many bedtime stories that won't get read.
Cobalt blue sunrises followed by fiery red sunsets,
The scale of privilege balanced with regrets.
My job pulls me here to Timbuktu, but
The heart of me is always with you.
Missed days and nights filled with great joy and laughter,
Too tight of schedules running here and thereafter,
To earn a living, son, this is what I do;
I miss every minute separated from you.
If a king's ransom I had, I'd be a stay-at-home dad,
Every night rubbing your head and tucking you into bed,
Long walks on the beach to take,
Seeing your smile when I wake.
So, son, keep in mind as you grow older,
A strong healthy body and oh so much bolder.
All those days and nights and days away too long,
I miss you every minute I'm gone.
If God should send me for that final flight west,
Don't be sad, don't protest.
Keep your head and body strong,
To win the challenge long.
And remember as you gander at the sky above you,
That forever in time I will always love you.

—Captain Rick Kerti

A Little Boy's Dream

"Tommy, what do you want to be
when you grow up?" the teacher asked
her daydream boy.
"I want to fly airplanes, ma'am.
I've loved them since I was three.
I drew my Grandma a picture of one . . .
just ask my brother Lee."
And the teacher smiled
and called on the next child,
keeping her thoughts to herself.
"What foolishness, to act like a bird!
Won't that little boy come down to earth."

But Tommy grew up
with his dreams trailing him
high into the sky.
He flies fighters for America,
so his countrymen won't have to die.
They don't call him Tommy anymore
when the canopy clamps shut
and the jet engine roar.
But when he checks his six
and grips the stick
and looks in the rear-view mirror,
on the face of his helmet visor–
The fighter pilot can see
the distant daydream boy
who'd wanted to fly
since he was three.

–Kathleen M. Rodgers

Dog Fight

War Eagles
SCREAMING across the sky
b
d m
i i
v l
e c

twist-turn
& circle
one another like
flying gladiators in a blue arena,
blasting and machine-gunning each other
until one is hit,
and like a wounded bird
f
a
l
l
s

crashing against the hard earth.

–John J. Kowalski

Why I Want to Be a Pilot

When I grow up I want to be a pilot because it's a fun job and easy to do. That's why there are so many pilots flying around these days.

Pilots don't need much school. They just have to learn to read numbers so they can read their instruments.

I guess they should be able to read a road map too...

Pilots should be brave so they wont get scared if it's foggy and they can't see, or if a wing or motor falls off...

Pilots have to have good eyes to see through the clouds, and they can't be afraid of thunder or lightning because they are much closer to them than we are.

The salary pilots make is another thing I like. They make more money than they know what to do with. This is because most people think that flying a plane is dangerous, except pilots don't because they know how easy it is.

I hope I don't get air-sick because I get car-sick and if I get air-sick I couldn't be a pilot and then I would have to go to work.

Author Unknown

Believed to be a South Carolina 5th Grader.

Wings to Fly

I was there once, as he pointed from the sky,
I was there once, with longing tears in my eyes–
Not so very long ago from runways end I would stare,
gazing at the planes as they approached, wishing that
I was there. –
I was there once, with the fever of the sky,
I was there once, but now I have my wings to fly. –
I was there once, where that young man quietly stares,
I was there once, saying those same prayers. –
I have not since lost the wonderment of what I saw,
Somehow longing anxiously within myself, pining impatiently,
yet respectfully in awe. –
Yes, I was there once, before I learned to fly,
and in time he will say pointing from the sky,
I too was there once, but now I have my wings to fly. –

–David A. Morris

First Things First

The boundary lamps were yellow blurs
Against the winter night
And I had checked the last ship in
And snapped the office light,
And paused a while to let the ghosts
Of bygone days and men
Roam down the skies of auld lang syne...
As one will now and then...
When fancy sent me company,
A red cheeked lad to stand
With questions gleaming in his eyes,
A model in his hand.

He may have been your boy, or mine,
I could not clearly see,
But there was no mistaking how
His eyes were questing me
For answers which all sons must have
Who build their toys in play
But pow'r them with valiant dreams
And fly them far away;
So down I sat with him beside
There in the dim lit shed
And with the ghosts of better men
To check on me, I said:

"I cannot tell you, sonny boy,
The future of this art,
But one thing I can show you, lad,
An old time pilot's heart;
And you may judge what flight may give
Or hold in store for you
By knowing how true pilots feel
About the work they do;

And only he who dedicates
His life to some ideal
Becomes as one with what he dreams
His future will reveal.

Not one of us whose wings are dust
Would call his bargain in,
Not one of us would welsh his part
To save his bloomin' skin,
Not one would wish to walk again
Unless allowed to throw
His heart into the thing he loved
And go as he would go:
Not one would change for gold or pow'r
Nor fun nor love nor fame
The part he played and price he paid
In making good the game.

And of the living . . . none, not one
Regrets the scars he bears,
The sheer uncertainty of plans,
The poverty he shares,
Remitted price for one mistake
That checks a bright career,
The shattered hopes, the scant rewards,
The future never clear:
And of the living . . . none, not one
Who truly loves the sky
Would trade a hundred earth bound hours
For one that he could fly.

If that sleek model in your hand
Which you have brought to me
Most represents the thing you love,
The thing you want to be,
Then you will fill your curly head

With knowledge, fact and lore,
For there is no short cut which leads
To aviation's door;
And only those whose zeal is proved
By patient toil and will
Shall ever have a part to play
Or have a place to fill."

And suddenly the lad was gone
On wings I could not hear,
But from afar off came his voice
In studied tones and clear,
A prophet's message simply told
For this is what he said
And why his hand will some day lead
Formations overhead,
"Who wants to fly has got to know:
Now two times two is four:
I got to learn the first things first!"
... I closed the hangar door.

–Gill Robb Wilson, 1938

The Broken Model

Will you have a place just here, sir?
Dusk falls soon these autumn nights:
We can watch the small ships hurry
To get home without their lights
And the big ones with the lanterns
In their wings ... like railroad trains:
Yes, I flown them all, I reckon:
Have they any future? Planes!
That's what you came here to ask me?

Oh, I see! Well! That explains!
Have an old time pilot coach you
How to answer when your son
Hands you proudly up the model
Of the aircraft he has done:
Feel you might direct his future
By the dreams that fill his mind:
But you have no childhood matching
Such experience in kind,
So your lips are sealed by loving
Lest the blind should lead the blind?

Well . . . this body scarce can answer
As these scars must testify
For I've blundered my full share, sir,
Through the years upon the sky
And great wealth is not the magnet
For the air is still frontier
And my name is only known, sir,
As an old time pioneer
Who will some day lose his rating
On the ships and disappear.

Yet I'd do the same again, sir,
Starve and freeze to hold my place
For the recompense is greater
Than appearances can trace,
So let's check against the crack-ups
Plus uncertainty and care
What might be the compensations
For the choice of this career,
If your little lad persists in
Launching models in the air.

Let us first consider friendship:
Life is great or small through friends:

Where one toils will most determine
Those with whom his future blends:
Men he finds upon the skyways
Will stir challenge in your son,
Teach him confidence and courage
For the thing that must be done,
And rejoice with him as brothers
When his spurs are fairly won.

Now, you'll want to know the danger?
Danger teaches men to walk
Careful with life's facts and figures,
Makes them quiet, slow to talk:
Flight is safe if safe men reckon
On its purpose and its plan,
Never discount danger's presence
But employ it as they can
Knowing that the great equation
Of the airways is the man.

Then there's contact with the grandeur
Which the man who flies will find:
Snow capped peaks and painted deserts
Canyons where white waters wind,
Mountains blue as powdered lapis,
Prairies gold as Spanish plate,
Tangled swamps and cyprus jungles
Where the wild fowl congregate
And the coral keys when sun sets
'cross the atolls in the strait.

And there's lure of distant places
Where no surface path is bent,
Lore of unknown lands and races
For the throttle and sextant:
Every pilot hears the voices

Calling, calling from afar
To the compass of his being
For the search of things that are
Waiting who has faith to follow
In his constancy, a star.

And there's solitude out yonder
Where a man can meet his soul,
Curse his follies, ease his heartache
Through a lonesome high patrol:
Nothing in your life can match it . . .
Flight at midnight 'neath the stars,
Plunge a thousand breathless meters,
Climb a'roar the moonlit stairs
Free o' fear and square with heaven
. . . God, the memories it bears!

–Gill Robb Wilson, 1938

Silly Old Baboon

There was a Baboon
Who, one afternoon,
Said, "I think I will fly to the sun."
So, with two great palms
Strapped to his arms,
He started his take-off run.
Mile after mile
He galloped in style
But never once left the ground.
"You're running too slow,"
Said a passing crow,
"Try reaching the speed of sound."
So he put on a spurt–

By God how it hurt!
The soles of his feet caught fire.
There were great clouds of steam
As he raced through a stream
But he still couldn't get any higher.
Racing on through the night
Both his knees caught alight
And smoke billowed out from his rear.
Quick to his aid
Came a fire brigade
Who chased him for over a year.
Many moons passed by.
Did Baboon ever fly?
Did he ever get to the sun?
I've just heard today
That he's well on his way!
He'll be passing through Acton at one.

–Unknown Author

Found on the Canadian Aviation Poetry Website (http://www.midwintercanada.com), developed by Stewart Midwinter, an active, internationally competitive hang glider pilot. He came upon this poem while riding on a train and talking to Terence Alan Milligan (born 16 April 1918). Milligan published the poem in his 1968 book, *A Book of Milliganimals*.

The Physical Sky

Downburst

I was a scholar of ocean and sky,
Who ought to know better than to rely,
On forecasts too early promising ease.
Weather grows fickle in still reckless breeze.
Troubled by doubt in this morning hour,
You held me and warned the day would sour.
"Subtropical ghosts cast spells of dull rain,
And may turn your efforts today in vain!"
But doldrums have left me feeble and bare,
From months aground molting limp in dead air.
Alas in white glass adventure takes flight,
Mindless of choices should lift lead to plight.
Aloft over mountains in Spring's embrace,
Strong thermals abound on the shear west face.
While darkened clouds begin to boil,
The miles pass by with little toil.
Shadows grow windward among vengeful breeze
Scattering warnings below in the trees.
Now far from home a trap has been born,
The intervals back are blocked by a storm.
Escape thru the veil of cloud crashing air,
Where reason is lost among rising terror.
Grasping for instinct as hail pounds the wings,
Emerging in mist among rainbow rings.
Crossing bleak mesas beneath blowing dust,
My energy burns away with each gust.
Downburst from the storm goes on without end,
"My God, I'm falling! Don't let me descend!"
Above on three sides, a box-canyon rim,
As tombstones reach up I cry out to Him,
"Save me dear Lord! Once again I have sinned.
The only way out seems into the wind."

Releasing control on faith I'm adrift,
Back into the rocks in search of lost lift.
There granite embers still glow with warm hope,
And spawn broken bubbles rising up-slope.
"Center one and get the hell out of here!"
It's time to look down on this place o fear.
Last moments of light are all that remain,
To final glide home and land in the rain...

–Dr. Scott A. Jenkins, 1987

Praying a Stall Won't Spin Us

Letting down through thunderstorms,
gusts like volcano blasts
slamming my head to the canopy,
I keep the needle centered,

boots toe-dancing the rudders, ready
to catch wings dipping and roll them
level. I watch the airspeed
bounce like a Geiger counter

and hold the nosecone constant,
praying a stall won't spin us.
Better jets than this have crashed
in lightning, wings snapped

and scattered. In unstable air
we find what dreams are made of,
why some birds hunt in water,
why eagles scorn the wide flat world.

–Walt McDonald

Van Gogh Sky

I am looking for the sky of Van Gogh here.
Boiling with energy.
Curvaceous in upheaval.
Vital.
Van Gogh could see thermals.
It is my great fortune and honored charge to feel them,
and to report the texture of the moment faithfully.
How then, can I not help but harbor this gratitude?

–Gary Osoba, February 1999

Rotor Winds

Rotor winds, I hate you,
You are no friend of mine.
You shake my plane,
You shake my frame,
You jar my very spine.

The only thing you're good for
And it's not much, I fear,
Is that when I'm tossing in your grasp
I know the WAVE is near.

I know that soon I'll leave you
For the peaceful, smooth ascent
And I'll soon forget
The bit of sweat
It cost me going through.

So you're nothing but a signpost,
"Rough Road" is what you say,

But on the other side
Is the rising tide,
So I'll travel by your way.

I'll take your up,
I'll take your down,
And I'll take it like a slave,
If you'll let me pass
To that air like glass
We call the "Cowley Wave".

–Tom Schollie

Thermal

You were as a feather lifted,
a soaring bird caught joyous in the thermal's core.
So not to rise in haste amidst the splendid journey,
you pulled away to the gentle currents at the edge,
only to glide in awe at new landscapes beneath.

Then, easing back to find the center's swift warmth,
You wondered who it was that moved?
Perhaps the thermal sought and found you too.

–Eric Stice, 1997, Eagle River, Alaska

Viewed From Another Angle

Grey the wind, grey the earth, grey the sky:
ragged nimbus fringes mist the view;
the engine's beat, turned back by earth and cloud
pulses round my brain; flying in a world of grey.

Patches of sepia light brown the ripening crop
and then extinguish. The sun's full circle,
paler than last evening's moon,
washes the level barley fields, then disappears.

The grey cathedral roof writhes, warps, tears,
and for an instant perforates; creating space
in which I spiral tightly upwards,
brushing against the breathing droplet walls of cloud.

I climb and climb past living cliffs, now black,
now grey, now white; bursting at last
into an arctic world, whose powerful sun
throws haloed shadows on the towering pack

of icy crystals, that filter colour out from light
still struggling down to earth.
Blue the sky, blinding the sun, brilliant the cloud,
that to those, earthbound, weeps down in shades of grey.

–*David Pedlow*

Winter High

One mile high and climbing;
and, before you ask–

No –
I'm not a member
of that club.

Not on this machine
where shrivelling cold
seeps almost imperceptibly beneath
the onion layers
of thermal packaging,

and gossamer threads of ice
catch the sun,
inching outwards
on the lines that hang
engine, seat and footrest
from the taught stretched parachute.

The terrier shake
of blue sky turbulence
bounces me against my straps,
swings me hard from side to side;

a pulse of pure adrenalin,
to keep me buzzing
long after
I am safely on the ground

–David Pedlow

Vapor Trails – 1

"Nothing but horror there," said the black man.
"Life but a missed glint, with few eyes watching,
and it is cold, cold."

"Stay," said the black man,
"Where I can shelter you and keep you warm."

The white pilot scoffed and touched
switches. His canopy shut out the rain.
His metal bird thrust him up through the clouds
toward the welcoming glare of the warming sun.

His shadow a circled cross, his wings
no longer pinioned by the bleak world's wax.
His thoughts only of all the spacebound things,
light-shrouded, flying free from the world's wrack.
Soaring out, out, his trail the comet's track,
seeking the light, he found only absolute black.

–John Clark Pratt

Small Planes and Dark Clouds

The sky was dark, the clouds were gray;
no small airplane should embark that way.
Foolishly our thoughts misplaced;
we had no idea of what we faced.
We took off with our flight plan signed,
hearts pounding, scared to death, into the blind.

The lightning snapped, the thunder roared;
white flashes zapped, too close we soared.

Our mouths were dry, our fingers tingled;
the clouds closed in, engulfed, we mingled
with gray on gray and black and white
we were in for some ride, like no other flight.

The controller called, we tried to respond;
but navigation equipment was well beyond
any realm of practical purpose.
Our only hope now was for a smooth landing surface
as our wings thickened with clear ice, not rime,
and we continued to fly on our imaginary line.

Given clearance to descend, we headed toward earth,
giving strikes of lightning a very wide berth.
On the approach much against our will,
we slowed the airplane, our hearts pounding still.
We needed less speed for our safety and protection;
We knew this from years of practice and direction.

Direction derived throughout the years,
from instructors and pilots with their own knowledge and fears.
We knew before flying to always check the weather,
which we did in depth, earlier, together.
We concluded this flight might be rough,
but felt it would be safe, though possibly tough.

Tough was the understatement of the year!
This flight taught us, through physical challenge and mental fear,
that even official weather reports occasionally reflect
that each and every pilot should have more respect...
respect for their own personal ability, and knowledge of the sky.
When clouds look THAT bad – don't fly!

–Patricia J. Rockwell

The Cloud

I see a high lonesome cloud in the sky.
He is small and a newcomer to the great heights
I will call it.
He drifts lonesomely along in the great heights.
He hardly sticks out from all the other clouds.
He is lost in his own territory
left at the mercy of the wind to take him away.
Now he is gone.
As senselessly as he appeared he has disappeared.
Swallowed up by the blue sky.
His brothers do not miss him any.
He might reappear as a big cloud but for now
his spot is forgotten in the great heights.
How senselessly to put a soul in the sky only to
make him die again.
Why…

–Daniel Hibbard, 1992

Sky Fever

I must go back to the sky again,
To the world of air smooth and soft,
And all I ask is a sleek ship
And a thermal to lift her aloft –
And the cu's kick and the wind's song
And the green ball hopping*,
Six thousand feet on a June day,
And the white clouds popping.
I must go back to the sky again,
For the call of the mountain wave
Is a wild call and a clear call

That lures the bold and the brave.
And all I ask is a west wind
And the cap cloud standing,
Twelve thousand feet o'er the mountain peak
And a ship that heeds my commanding.
I must go back to the sky again,
To a soaring nomad's life,
To the hawk's way and the eagle's way
Far from the daily strife.
And all I ask is a street of cu
'Til the long trek is over,
And a gentle glide at the set of sun
To a soft field of clover.

–Author Unknown

This poem is a take off from Masefield's "Sea Fever."

The Beam

Rest easy back there!
We are riding the beam!
A ribbon of sound,
A mechanical dream
That pulses and throbs
In the dome of creation,
Forming our highway
To each destination,
Along which we drone
In vast isolation.

Rest easy back there!
We are riding the beam!
High over the storm

Where the lightnings gleam
Through Stygian night
Or in deep overcast,
We check by the throbbing
The leagues that are passed,
And hold true our course
By the radio mast.

–Gill Robb Wilson

Those Who Make It Happen

The Eagles and Black Nights

In days of old when knights were bold and Arthur was the king,
The table of round protected the crown,
and Merlin did magical things.

He came to me in a cloud of mist and took me on a flight
that led me through the hall of time to a
world of power and might.

Two nations ruled this future world, it was very plain to see,
but the land of torch and lady was by far the best to me.

The nation of red sought dominion with
Badgers, Bears and Foxbats,
but the nation of blue was valiant and true
and defended by Eagles and Tomcats.

Most powerful of all was the Eagle,
and his work was never done,
but he always proved himself the best in the land of the
midnight sun.

To the North on an isle of ice and fire, an Eagles nest did stand.
Dauntless were these birds of prey
for they feared no beast nor man.

These birds of prey, all silver and gray,
carried death on talons and wings.
They fought for the freedom and rights of man,
not communist, country, nor king.

Their eyes were clear, their power great,
they looked through clouds and night.
Finned vessels hung beneath their wings
and saw with invisible light.

The vessels were marked with sevens and nines,
and fire followed their flight,
and when they truck a Badger or Bear,
they exploded with a terrible might.

Black nights, both men and women, did tend their every need.
Their names were called "Maintainers,"
and they had an honorable creed:
to keep the Eagles ready,
to protect the skies on high,
so freedom rings forever above
their country's skies.

–Author Unknown

The Forgotten Man

Through the history of world aviation
Many names have come to the fore
Great deeds of the past in our memory will last
As they're joined by more and more.
When man first started his labor
In his quest to conquer the sky
He was designer, mechanic, and pilot,
And he built a machine that would fly.
The pilot was everyone's hero.
He was brave, he was bold, he was grand,
As he stood by his battered old bi-plane
With his goggles and helmet in hand.
To be sure, these pilots all earned it,
To fly then you had to have guts.
And they blazed their names in the Hall of Fame
On wings with bailing wire struts.
But for each of our flying heroes

There were thousands of little renown.
And these were the men who worked on the planes
But kept their feet on the ground.
We all know the name of Lindbergh,
And we've read of his flight into fame.
But think, if you can, of his maintenance man,
Can you remember his name?
And think of our wartime heroes,
Smith, Graham and Ernst.
Can you tell me the names of their crew chiefs?
A thousand to one you cannot.
Now, pilots are highly trained people
And wings are not easily won.
But without the work of the maintenance man
Our pilots would march with a gun.
So when you see the mighty aircraft
As they mark their path through the air,
The grease stained man with the wrench in his hand
Is the man who put them there.

–Author Unknown

For the Guys on the Ground

For all those winged aggressive types
It's time to set the story straight
Oh, your glory is unquestionable
But it's time to clear the slate
The target was destroyed
You know your glory found
As you take your many praises
Remember the guys on the ground
They take your constant prodding
And never blink an eye

They endure the worst conditions
All that you may fly
No parts no time or money
Their heads will never swell
No matter how they're treated
The mission will not fail
Your safety is their prime concern
Their hands control your fate
Regardless of the obstacles
The mission won't be late
If there is a ground abort
Or things don't go your way
You can rest assured
It was an Ops delay
We know you're on the pointy edge
The ones who keep us free
But if you don't accept the shaft
Leaders you'll not be
Next time you break the surly bonds
And dance on sun-lit skies
Consider those who put you there
And thank the maintenance guys

–Dave Ray

Just Another Flying Day

The rising sun, across the tail.
The long cold night, cracks its shell.
The morning chill, it's here to stay.
It's just another flying day.
As most arise, at the crack of dawn.
The first of many flights are gone.
The nights as cold, as the day is long.

Not to worry, their will is strong.
Their hearts are gold, their intentions pure.
Take your best, they will endure.
The shifts half done, as the first returns.
No time to rest, they're all quick turns.
The pace is fast, no time to rest.
Not to worry, they're the best.
They send them on, with sharp salutes.
All professionals, no disputes.
No time to boast, as the sun droops low.
The aircraft land in constant flow.
They give it all, and then some more.
They're whom we ask to fight the war.
Darkness comes and steals the light.
Inspect, repair, we fly at night.
They step inside, to warm their hands.
In some far off place, in foreign lands.
On the line, they work the spare.
The lights are bright, with blinding glare.
Mission first, without concern.
Today, their pay, they know they'll earn.
Tired, cold, and worn with pride.
Thank the Lord, they're on our side.
Their shift is done, they choose to stay.
It's just another flying day.

–Dave Ray

The Line Mechanic

The pay is not the best
The schedule not the greatest
But I chose to take the test

And now I'm different than the rest
I work night nights to keep them flying
Sometimes till the morning light
It seems I'll never win the fight
But I will always keep on trying
I choose to sacrifice my life
So they can make their destinations
I choose to miss my wife
So I can feel all the sensations
The pride I feel as they rotate
And climb to bluer skies above
The joy I feel is sometimes great
It's like a different kind of love
When the pilot says she's broke
And I get her back on line
I cannot take it as a joke
When it could be his life or mine
I worked hard to get my ticket
Worked days and studied nights
I am glad that I stuck with it
As I walk around and check the lights

–Hansel Herrera

The Nameless Legion

When the records tell the story of the conquest of the air
By the instruments and motors and the ships,
Of pioneer, and Bendix, with their gauges and their breaks,
Of Wright, and Pratt & Whitney, and Eclipse,
Of Boeing, and of Douglas, and the salty Thompson valves,
Of Kollsman, and of Sperry, and of Beech,
Of technicians and designers, and of schools and engineers,

And have catalogued the skill and scope of each,
Will they swing a shredded wind-sock
in the west' ring winds o' time
For the nameless men who toiled behind the scene
With the blow torch and the pliers, with the pistons and the plugs
That the glory might be as the glory's been?
When they tell of Stearman, Stinson and Bellanca, Ryan, Breeze,
Of Sikorski and of Loening with their boats,
Of Fairchild with the cam'ra, and of Hamilton and props,
And of Edo and the landings made on floats,
Of Rosendahl and airships, and of Goodyear and balloons,
Of Martin, and Menasco, and of Vought,
And describe how each was loyal to the single thing he loved,
And the many gallant battles that he fought,
Will their flight plan think to mention some citation to the years
For the nameless ones who kept the vision too
By the faith in which they fashioned every spar and rib and strut,
That the game might have its chance at coming through?
When they dwell on Trippe and Lockheed
and on Waco, Cessna, Ford,
And on Pitcairn, Warner, Jacobs, Grumman, Fleet,
When they eulogize Aeroncas and the Rearwins and the Cubs,
And the ones with which the little ships compete;
Will they throw the switch o' mem'ry
for mechanics they have known,
For the greasy tribe which keep the game in trim,
The patient lads who scan the skies until the ships return –
Will they spin the prop in praise of such as him;
Will they wipe their hands in fancy on the coverall's 'e wore
Though the records do not rate him by his name,
When they catalogue the backers and the pioneers of flight
In the annals of the men who made the game?

–Gill Robb Wilson

The Parable of Joe

Let us consider the ground crew,
Too often forgotten by all.
They get to do time out on every flight line
But never to carry the ball.

Joe is a chap who is needed,
A problem that won't go away,
But remember of course that the knight on his horse
Would tell you the same in his day.

He'd say he was lacking in armour,
And his new iron pants weren't quite right,
And did he show pity to the overworked smithy?
You can't expect that from a knight!

"I must have a bigger brick privy,
Ye drawbridge is terribly short,
And get me a steed with a little more speed,
My charges must never abort!"

"Don't tell me thee can't find the money,
Those problems don't move me at all.
Why, I've got a notion to block thy promotion–
Get snappin' and get on ye ball!"

So the vassals and serfs got to sweating
And bending their backs a bit more,
'Cause it wasn't the rage in the chivalrous age
To ask the lord why or what for.

But suddenly – horrors of horrors!
The knight was knocked from his mount,
Pierced to the marrow by a little ol' arrow,
And down went m'lord for the count.

There lay the lord and the master,
Flat on his back on the field,
And he yelled and he howled that he must have been fouled
And he swore he'd remount ere he'd yield.

Well, sure enough, centuries later,
A couple of vassals named Wright
Glued a few things to a couple of wings,
And handed it all to the knight.

Up into the cockpit he vaulted
And tried on the saddle for size.
With throttle full bore and a rush and a roar,
He tore a few holes in the skies.

But the Joes were back where they started
And they put down their tools with a sigh,

'Cause they knew sure as fate
When he landed the crate,
They'd have to perform the DI! *

"Build me another big hangar,
I need one more mile to take off.
This aircraft won't do, I must have Mach 2.
Attend to it will you, old toff?"

So the chargers grew bigger and faster,
They belched out their fire and smoke.
To the knight it was pleasant – but not to the peasant,
Joe could never savour the joke.

Then up and spoke an old boffin,
He of the rapid slide rule,
"I have in my pocket the plans for a rocket,
I'm telling you, knight, it's real cool!"

"It's almost as big as a mountain,
With cockpits and saddles galore."
(Now surely by rights we should fill it with knights,
And we shall be bothered no more!)

Now after all was assembled
And the brass gathered 'round for a look,
You could tell by their sighs and the gleam in their eyes,
They were ready to swallow the hook.

Into the rocket they clambered,
Each to his own private place,
And eager as beavers they played with the levers
'Til the monster roared off into space! . . .

Thus the Joes did the old world inherit,
Mountains and river and plain,
While the knights in the sky go hurtling by
As they circle the sun once again.

–Author Unknown

*Joe=Tinker or mechanic, DI=Daily Inspection, Boffin = British slang for a scientist

Historical Feats

of Wrights, Lindbergh, and Earhart

Wright Field, 1944: The Ready Room

On the benches in this room were morning vacant spaces
For some that saw the sky turn dark, at noon, with youthful faces.
Too soon, too low, too high, too fast
They hurried on, into the past.

The crashes often most were blamed on grievous pilot errors,
But on the hill where the coffin was, six pilots were the bearers.
The feathering button that didn't work,
The gear-down switch with the funny quirk.

Propthunder at night when the throttles were tight
On take-off, while one ran away – away.
"Help on the wheel, turn up the blower,
"Left rudder tab," while the pressure sank lower.

Flinders and flames, flinders and flames,
The names soon forgotten, along with the blames.
The railway track on runway naught,
The lives of friends that were dearly bought.

The air was rare and the cold got old
At forty-one thousand or better.
"Look out the window – that seems new.
"How strange, the edges of space turned blue."

The sands in their glasses dwindled by grains
While they peed in the bomb bays of redlined planes.
God bless Pratt and Whitney and Wrights'
And pray for doomed ghosts in aluminum kites.

Listen. The echoes ring here yet, of engine test cells roaring.
Above drift the skies they touched

and tried, alone, aloft, exploring.
The games of chess unfinished still –
The rain, the line, the morning chill.

–T. P. Leary, 1981

Old Pilots in the Crowd at Kitty Hawk

The guide says the mannequin with the mustache,
that's Orville. The guide's eyes twinkle
and he points. Orville's hanging head-first
into history, his left fist stiff on the controls.

Inside the visitors' center, the mock-up
never flies, just what scoffers always warned
those quacks from Ohio. Fake props are plastic,
the faces painted gaudy as corpses,

hard to feel for dummies in crash tests.
In grainy pictures, the brothers never smile,
but both look vulnerable and real. Wilbur,
we know, died young. The guide asks pilots

to raise our hands, and all who do
are old. We've all flown fighters, transports,
bombers that could wipe Kitty Hawk off the map,
turn the sandy Outer Banks to glass.

The guide's dramatic, grabs the wing
and twists, proving the skills of lift
and turn. We'd rather he step aside, let us roll
that old crate outside, flip for the right

to go first on our bellies. Goggles down,
cracked leather gloves back on, we'd nod,
face to the wind and history be damned,
full throttle at thirty miles an hour,

a hundred, eight hundred feet down range.
What was Wilbur's record that day, almost a minute?
Never mind the hedge of trees they've planted
a hundred feet beyond, watch this.

—Walt McDonald

We

It would have been as fine and brave a deed
Had he who did it been worn grim by time,
Swollen with vanity or sharp with greed,
Or tarnished by life's grime.

It was not just his courage, nor his high
Accomplishment, nor all the hopes unlocked
For future years, that wakened such a cry
It seemed the whole world rocked.

Youth . . . it was Youth we cheered around
the earth,
A knightly soul and body strong and clean,
Unstudied courtesy and native worth,
A modesty serene
Through storms of praise that well might over-
bear

A mind less anchored to eternities.
More than the glorious conquest of the air
We cheered such things as these.

He has youth's answer to the wail of those
Who mourn a world swift crumbling to decay.
He was the hope that every woman knows
But some may never say...

The son who died, or was never born,
Who might have been like this. He was the
bright
Indomitable breaking of the morn
After a weary night.

And now, as all tides must, this tide will fall.
As other names fill the world's mouth, will he
Sign with a wistful longing to recall
His day of deity?

No. Of his chosen labor's faithful round
That day was but an arc. There was no need
For him to shift life's centre when he crowned
The vision with the deed.

The wings that bore him were no new, untried
Experiment, for one great venture spread,
Nor were the dreams his daring verified.
They were his daily bread.

He and his wings are one, soaring above
The silence and the praise of circumstance.
O happy man! His duty is his love,
His work is his romance.

–Amelia Josephine Burr

Lindbergh Flies Alone

Alone he soars, this Eagle of our time, –
Intrepid, unknown youth with steadfast eyes,
Into the untracked distance of the skies;
Through mists of mourning on a quest sublime.
In non-stop flight, as none has ever flown,
He flies alone.

Alone, yet companied by courage high –
With faith, and skill and understanding true;
One with his plane, like bird across the blue, –
His purpose firm, to conquer or to die.
Still, through it all, there thrills the undertone,
He flies alone.

And all the way, as throbbing soul to soul,
The heart of multitudes outpours in fervent prayer
That dauntless courage, speeding thus through air,
May find its haven, reach its longed-for goal.
While countless prayers ascend to Heaven's throne,
He flies alone.

He reaches Paris, – wins the accolade;
Ambassador of air, brings to the scene
New bonds of friendship; and with modest mien
Receives high honors to his prowess paid.
Into his Country's heart, – well-loved and known,
He flies alone.

–Cornelia Fulton Crary

Lindbergh

Did you feel our hands on your hands as you guided
The lone bird over the sea?
Blonde young Viking, we were with you
Bearing you company.

But it was you who chose the danger,
You who took the chance,
You who tossed your life like a coin, lad,
To bind your land to France.

You have made your mother our well-loved mother,
You have made your smile our smile;
You have joined the nations brother to brother,
You have brought back peace for a while.

Blonde young Viking, flying, flying,
Like a sword that breaks the blue,
While the world remembers the men who made it,
It shall remember you.

–Mary Carolyn Davies

A Faithless Generation Asked a Sign

A FAITHLESS generation asked a sign,
Some fresh and flaming proof of human worth,
Since youth could find no flavor in life's wine,
And there were no more giants in the earth.
Then out of the gray obscurity he came
To laugh at space and thrust aside its bars;
To manifest the littleness of fame
To one who has companioned with the stars.

The drought of greed is broken, – fruitful streams
Of courage flow through fields long parched and dead;
Young men see visions now, old men dream dreams,
A world moves forward with uplifted head:
A lad with wings to dare had faith to rise
And carve proud arcs across uncharted skies.
What shall they say of him who add his name
To the proud page of earth's remembered ones?
Was it loneness of his flight to fame
That placed him with the well beloved sons
Of all the ages? Rather let them say,
"Because his heart was greater than his deed,"
"His courage higher than his cloud-strewn way,"
"His faith more shining than his silver steed,"
"Men squared their shoulders when they spoke of him,"
"The lad who brought from solitude or star"
"A splendor wealth and homage could not dim,"
"To glow in fame's fierce light without a scar:
"A secret every man of them could share, –"
"All things are his whose dreams have wings to dare!"

–Molly Anderson Haley

A Chantey for Celestial Vikings

Swallows swarm
Among the wind-blown blossoms.
Swans cleave the foam-clovered sea
And leave a caressing furrow
In the meadowy surf.
The eagle flies high
And alone.

I, too, have flown
As the eagle flies.
I have seen quilted islands,
Polka-dotted with orchards,
And spotted with lakes
Like staring, embroidered eyes.
I have seen the swan-path
And the seal-bath
Dangle like crystals
From my cloud-cooled wrists.
I have seen furrows plowed by whales
Flutter behind me like tattered silks,
A tail for my silver kite.
I have outwhistled gales
And playfully patted the hounds of day
As they barked at the heels of night.

I have outflown the gulls
And the clouds.
I have companioned the sun
In its singing flight over ocean
And watched it drop like a rocket
Into the starry shallows.
I have felt fogs close around me
Like clutching fingers.
I have heard the death-tinkle of sleet
On tomb-white wings
And the vaporous gibberings
Of sea-weary ghosts.
I have brushed their frozen breath
From my rime-wet cheeks
And laughed at their sinister warnings.
I have blown the mists away
Like smoke-rings from a giant's pipe
Or darted through them
Into a skyful of stars

A skyway without horizons.
I have vanquished the ravenous shadows
And found flares of furry luminance
Beyond the deepest dark.
I cannot die
Who have flown as eagles fly
Into the blue unknown.
I have throttled oblivion.
I have fathomed the myths
Of space and time.
One to me are height and depth,
Daylight and dark.
I have fronted the universe
And found it friendly.
I have broken its stormy rhythms
With my rainbow laughter.
I have shattered its terse geometry
And its brittle syllogisms
With an arrogant gesture.
Hereafter
There will be rents in its seamless vesture.
It will bear the illogical scar of me
Through the far recesses
Where eagles fly
And the plangent wildernesses
Of sea
And sky.

I am one with immensity.
I cannot die.

–*Earl Marlatt*

The Hour

SO a man's moment falls.
Across his long-kept vigil suddenly,
At edge of some chill midnight, like a cry
The lifting bugle calls.

Into the gloom he came,
Mist on his forehead, eyes alert, aware,
His life's old resolution burning bare,
Tapered to torch-like flame.

Off – he is off and gone!
Out where a wrath of storm and cloud is weaving,
Climbing the stairway of the night, and cleaving
The stark, astounded dawn.

Against his going hurled
The winds of all the everlasting years,
And dark and death upon his trail, he clears
The chasm of the world.

So, of its own strange power,
A spirit's dream draws true. Heart-shaken, we mark
Across an old earth, sounding in the dark,
The strokes of a new hour.

–Nancy Byrd Turner

Amelia

Somewhere a fin on a lazy sea
And a broken prop on a coral key,
Somewhere a dawn whose morning star
Must etch dim light on a broken spar,
Somewhere a twilight that cannot go
Till it kisses the surf with afterglow;
But here, only silence and weary eyes
And an empty hangar and empty skies.

Somewhere the toss of a tousled head
In the secret of the angels overhead,
Somewhere a smile that would never fade
As the score reversed in the game she played,
Somewhere a spirit whose course held true
To do the thing that it wished to do;
But here, only silence and weary eyes
And an empty hangar and empty skies.

–Gill Robb Wilson, 1938

Pushing the Envelope

FLIGHT TEST

Test Pilot

So long as this is a free man's world
Somebody has to lead.
Somebody has to carry the ball
In a word and thought and deed.
Somebody's got to knock on doors
Which never have known a key.
Somebody's got to see the things
That the throng would never see.

Hotter than thrust when the boost is hit
Somebody's faith must burn.
And faster than Mach when the rocket's lit
Somebody's mind must turn.
Somebody's got to get the proof
For what the designers plan;
And test the dreams that the prophets dream
In behalf of their fellow man.

Somebody's got to think of pay
In terms that are more than gold
And, somebody has to spend himself
To buy what the Heavens hold.
Somebody's got to leave the crowd
And walk with his fears alone.
Somebody's got to accept the thorns
And weave for himself a crown.

It is ever thus as the ages roll,
And the record's written clear.
Somebody has to give himself
As the price of each frontier.

Somebody has to take a course,
And climb to a rendezvous
Where lonesome man with a will to learn
Can make the truth shine through.

–Gill Robb Wilson

What Is a Test Pilot?

When you think of it,
the history of human flight is no more than
a scratch on time's surface.
But the moment has been magical.
The ability to fly has transformed man from
a struggling earthbound biped...
limited by geographical circumstances
to the barest acquaintance with most of his fellow men...
into a winged creature.
Into a being who, as the poet said, has
"slipped the surly bonds of earth
and danced the skies on laughter-silvered wings."
Into a soaring, questing adventurer...
of who knows what potential.

Flight has re-cast nations,
re-made civilizations,
reconstructed our view of the earth itself.
Now the very universe ... its threshold already breached
Awaits man and his wonderful flying machines.
The dreamers and the thinkers began it all.
The improbably patent-seekers, with their silly gadgets
and their funny contractions.
The men who were laughed at.
They started it.

But it took another kind of man to make it possible,
to prove it could be done.

He is the test pilot.
His breed is rare, his number few.
He really is not one kind of man.
But a distillation of many kinds of men.
The best kinds of men.
Strong men, brave men, wise men, curious men.
Like the dreamers and the thinkers, he too is a visionary.
But he also is a realist, and as mentally
and physically tough as men come.
He is a man in love with life because life is
the essence of what he is doing.
But he is a man willing to risk death–
because he believes in what he is doing.
A man who, again as the poet said, has
"trod the high untrespassed sanctity of space,
put out my hand and touched the face of God."
Above all,
the test pilot is a persevering man,
smart and courageous enough to accept
temporary failure for what it is–
a pebble beneath the steamroller of progress
and with sheer determination,
grind each setback into mealy dust.
Some of the names you know better than
your next-door neighbor's;
some already are bigger than legend.
Some are faintly familiar; some you have never heard before.
But each is, or has been, a man with a purpose:

To fly where others
have not, cannot, dare not.
To push past barriers thought by all but a few to be impossible.
Because of men like them,

such barriers really never exist.
They never will,
as long one of their kind still lives.

And as test pilots, they share ... or shared ... certain
common qualities:
caring, ingenuity, strength ... physical, mental and
psychological ... and determination.
But each is an individual. No two are alike.
After each was born, the mold was smashed.
The least of the test pilots is a storybook character,
the giants among them the stuff of folklore.
What makes such a man?
Who is he?
Why is he?
Where does he come by his resolve?
His myriad talents? His inspiration?
What is he like as a person? A friend? A husband?
Where does he go tomorrow?
And, then, who follows him?
Farther and farther.
Past the stars, past the moon–
past the rim of the known?

–Author Unknown

Read at the Nineteenth Awards Banquet for the Society of Experimental Test Pilots, 27 September 1975. Found in the historical archive files at the United States Air Force Museum at Wright-Patterson Air Force Base in Ohio.

Test Pilot

Ten thousand grand for a dozen "G's"!
(And no one can take thirteen of these)
And all I do for this princely dough
Is climb as high as the ship will go
And dive till the needle hits the pin
In the gadget that shows how fast I've been,
Then pull the stick to reverse control
To see if I and the ship stay whole?

If the wings stay on and fittings tight
And the elevators function right
And ailerons don't flutter away
Or pressures buckle where pressures may,
I climb again with a like intent
To prove it was not an accident
And pick up "G's" with a spinning fall
To see if she's weak that way at all.

I loop for balance and stall for glide
And whipstall hard with the flaps out wide:
I spin wheels up and I spin wheels out
And spiral and slip and slam about
In landings rough as a tank could stand
To prove each thing the designer planned:
Then I taxi into an open ditch
And don't get paid for the son of a bitch!

–Gill Robb Wilson

840 by Accident

Men of science,
Drab and dreary,
NUTS to all your fancy theory;
Laws you've made are all archaic,
One might even say prosaic;
THUNDERBOLTS express defiance,
For the current forms of science;
Other aeroplanes are bound,
Such as by the speed of sound;
THUNDERBOLTS, which never heed it,
rather easily exceed it.

Past the range where meters read,
Pilots have to guess the speed,
Pilots' guesses can't be fairer,
We've a way that's free from error:
Hold the stick between your knees;
Hang your fanny in the breeze;
When your fanny starts to smolder,
You're the newest record holder.
We can prove by calculation
Any pilot's estimation.

If our advertising's right,
soon we'll beat the speed of light.

–Author Unknown

ASTRONAUTS AND SPACE

The Leap

They leaped from myths and dreams
From cliffs and sand dunes,
To the edge of space, the lunar surface.
Curiously, painfully they leaped
In the slow motion of centuries of desire.
Icarus, Da Vinci, the Wright brothers:
Giants who dared leap beyond the horizon.
Armstrong . . . but a footprint of history;
The last step, the first step,
A mortal once again thrashing the skeptics
And putting truth in the mouths of soothsayers.

–Preston F. Kirk

To a Husband Who Must Seek the Stars

In your eyes the first glad token
As when first our love we proved,
So your mind to mine has spoken
Just as if your lips had moved.

You are saying. . . yes, I know. . .
That the lure of space beguiles.
You are pleading. . . "Let me go,"
Not unwilling, but with smiles.

Can you love me, and still choose
Whispers that I cannot hear?
Late to love, how can I bear to lose
Content for some inconstant sphere?

Tell me how you see my role...
To stay, to wait, yet yearn to go.
Where is the comfort for my soul?
You, my love, have helped me know:

I'll be unafraid, undaunted.
Yes, of course! I need not face
Any peril; or be haunted
By the hazards you embrace.

I could have sought by wit or wile
Your bright dream to dim. And yet
If I'd swayed you with a smile
My reward would be regret.

So, for once, you shall not hear
Of the tears, unbidden, welling;
Or the nighttime stabs of fear.
These, this time, are not for telling.

–Pat Collins

"Mountain"

July 20, 1969
Eagle U.S.S. Hornet
Columbia CAPCOM
Armstrong-Aldrin MOL-.
Apollo-Collins Cancel
Borman Apollo 12
Apollo 8-Earthrise Mission Control
Deke Slayton "Sea of Storms"
Chris Kraft Budget
Apollo 1 –FIRE Apollo 13 –Explosion
Saturn V Grissom Congress – billions of dollars
National Unity White Apollo 14 - 17
Houston-Vostok Chaffee What to do?
Gemini-Surveyor Mourning Pause............
Telstar - Von Braun Fewer Jobs
integrated circuit Skylab
ICBM 85 Days
Echo - Agena Cape Canaveral Weeds
Shephard - New Frontier Pause ....Quiet
Gagarin Worthless frontier
Mercury Space Shuttle (?)
Cape Canaveral – Maybe
Goal Quiet
Kennedy Silence
NASA VOID!
Explorer
Space Age
Education
Science Why build a mountain
Technology to tear it down?
Sputnik
October 4, 1957

–Daniel C. McCorry, Jr., 1975

A Nation Cried

A nation cried today
For seven of her children are gone
She had watched them grow,
Listened to their hopes and dreams,
And helped them to achieve them.
The suddenly ... A FORCE
Stronger than she,
Took her children's lives
Today the country weeps
For they had all seen the flames
As they shot across the horizon,
Heard the explosion,
Echo across the countryside.
And cried,
For their perished brothers and sisters.
They stood in shock for forty-five minutes
As pieces of the celebrated shuttle
Fell slowly back to earth.
The disbelief that had so long
Shown in their eyes ... faded...
Faded into the knowledge
Of what had just happened.
The shuttle was gone.
And seven brothers and sisters with it.
The nation would recover,
She would continue,
But the memories of her children
Would never disappear, never fade.
For she would remember them always
In the spirit of courage and adventure,
They had shown as they boarded
The magnificent space shuttle "Challenger."

–Misty Sorensen "The Mystical One"

AIR RACING

Race Gear

A saddle on a motor
Burnin' dynamite for gas,
As little liftin' surface
As will hike it off the grass,
A thousand roweled horses
With a feather for a girth
Three hundred miles an hour
Fifty feet above the earth!

The breed o' man who rides 'em
Is an optimistic guy
With magic in his fingers
And a telescopic eye,
A throttle bendin' genius
With 'is neck upon 'is nose,
His nervous system sweetened
With intestinal repose.

An autumn day of shadows
With the wind across the lake,
A bonus for a record
And a fortune for a stake;
But hell is in the makin'
And the devil sets the pace
When they tangle out at Cleveland
In the Thompson Trophy Race!

–Gill Robb Wilson, 1938

Reno Races

Gather 'round me, younger pilots,
Gather 'round me, future aces,
Gather 'round me, little children
With bright eyes and shiny faces.
Sit beside me, pretty damsels,
Dressed in satin, dressed in laces;
And I'll tell you many stories
Of the epic Reno Races.

But I must go and I must rest now,
For I am old, and I am weary.
I have flown too many winters,
Over oceans dark and dreary.
Seen too many summer squall lines,
Seen their lightning, green and eerie
Seen too many mornings' sunrise
Through sleepless eyes bloodshot and bleary.

But come wake me in the morrow
When the skies are clear and sunny,
When the Warbirds sing their music,
When their magic fills the air!
With no limits to their power,
With no limits to their speeds;
Then I'll tell you more of Reno,
And of Racing pilots' deeds.

Of White Knights in shining armor,
Riding fire-breathing stallions;
Of the battles I have witnessed
In the desert north of Reno.

Of fearless aviators dying,
Of fearless men who keep on flying,
Men whose fortunes have been sold,
In their quest for Reno's Gold.

Climbing high to meet the challenge,
Joining up in stacked formation;
Like an angry swarm of hornets
Down the "chute" with engines whining
Red Baron's jet has set the pace:
"Gentlemen, you have a race!"
For the fans, it's most exciting,
Arc of smoke from red jet rising!

Now past the pylons, blurred and flashing
Through the Vale of Speed they're dashing,
'Round number seven speeds are building,
Pressures in the engines mounting!
Past the Starting Flag at Reno,
In September desert glow,
Past the unseen crowds assembled
Cheering madly, shouting "GO"!

"Go Tsunami! Go you Strega!"
"Go White Lightnin'! Go Rare Bear!"
Go you modern gladiators,
Fly you swiftly through the air!
Fly as fast as you would care to,
Fly as low as you may dare,
'Til the Checkered Flag at Reno
Signals "VICTORY" in the air.

Climb your mount now to the heavens,
Engine cooling, killing speed;
Checking gauges, checking pressures,

Checking damage to your steed.
Gear and flaps now coming downward,
Concrete rising to your wheels,
Kissing earth, canopy opens,
Cool fresh air, how sweet it feels!

Taxi inbound, taxi slowly,
Taxi proudly past the stands.
People clapping, people cheering,
Wave now, smiling to your fans.
To the pits, dismount, relaxing,
Glowing in the setting sun.
This year's work is nearly over,
Reno's toil is nearly done.

Mortal men now pay you homage,
Mortal men, who do not fly.
Take their trophies, take their tribute;
This shall pass.... and then..."Good-bye."
But your names shall be immortal,
And your fame will never die.
Modern Knights in shining armor,
Riding stallions in the sky!

–Michael J. Larkin

Air Race

The engine started, 1-2-3;
we rushed to take off anxiously.
Up off the runway, into the blue;
our senses keen, our course was true.

Beautiful day–crisp and clear–
until some clouds began to appear.
With heat and humidity they were fed;
soon lightening was seen overhead.

Cloud bottoms came down very low;
some of the hilltops didn't show.
Several routes we couldn't use,
so around the hills we quickly cruised.

The rains came on, the thunder boomed;
and up ahead the mountains loomed.
We passed each airport, did our fly-by;
watched the storm with an eagle eye.

But as this was our first race,
we weren't too sure about the pace.
Should we fly slow to save our fuel,
or was "full speed ahead" the rule?

We opted for speed; everything looked fine
on our way to that finish line.
Ten miles out we called the tower;
poured on the coal, still more power.

Circled the boundary, checked our time;
did the final fly-by; felt sublime.
We landed, taxied to our spot.
The air was cool, our cockpit hot.

Cheers filled the air. We felt elated,
dreaming of trophies we thought we rated.
But now to refuel: the gas pumps gushed.
The judges measured. Our voices hushed.

'Twas then we knew from our airplane's thirst
there was no way we could be first.
We were way over the prescribed limit;
we knew next time we'd have to trim it.

That night when prizes were all passed out,
we realized what air racing's really about.
Just flying, for us, won't be quite the same.
Air racing, we've found, is the name of the game!

–Patricia Rockwell

The Bendix

Stabbin' exhausts in the morning,
Hangars at Burbank a' glow,
Fog rollin' in from the ocean,
Mechanics warmin' 'em slow,
Everyone wishing each other
Luck for the get-away climb,
And this is the race o' races
In speed and distance and time!

Spannin' the continent east'ard,
A' raisin' the Jersey shore,
Take all the gas you can carry,
Wish you could carry some more,
Ponder the tape for the latest
Temperature, dew point and glass,
Learnin' its instrument flyin',
Wonderin' when it will pass!

Always there's weather to battle,
Thunderheads crownin' the peaks,
Satan presides in the cockpit,
Static and headache and leaks,
Wind changes no one can tell you,
No place to land if she cuts,
And what you need for the Bendix
Is inexhaustible "guts!"

Nobody's luck of a minute
Can hurdle the nation's chest!
Only by bringing together
Elements worthy o' test,
The pilot, ship and the motor,
Radio, compass and fuel
Blended to finish or perish
Seeking the ultimate goal!

–*Gill Robb Wilson*

Death by Flying

The Troop Who Rode One In

We should all bear one thing
in mind when we talk about
a troop who rode one in.

He called upon the sum of all his
knowledge and made a judgment.
He believed in it so strongly that he
knowingly bet his life on it.

That he was mistaken in his judgment
is a tragedy . . . not stupidity.
Every supervisor and contemporary
who ever spoke to him had an
opportunity to influence his judgment.

. . . so a little bit of all of us goes
in with every troop we lose.

–Author Unknown

To the Man I Never Met

Everything has changed. It's true:
the world forgets. I have
remnants of your voice, the harsh
resonance of memory. You would laugh,
I know, at how I first began to bridge
that distance. How I slid off the runway
in a T-37, the kiss of black tires
on burning pavement. How
I shit myself the first time
I saw a spin: a spiral falling

at twenty thousand feet a minute,
as if the earth had grabbed
with two hungry talons, wanting
to pull me deeper
into its body. But then
the time of my first night solo:
I cut the lights in the cockpit
and hung in the shadow
of the moon's breath, the chill
when I saw my helmet visor
reflected on the canopy at forty
thousand feet.
            I think of Brian,
his plane coming apart
in the jungles of that sad country,
mask melting his skin:
graft of black flesh, a human voice.
I think of Scott: failed parachute, earth
swelling beneath him like a net.
I think of Peter: the final turn
he never made. I think of Gus:
whine of engines saying *'something's
wrong'* as he lifts off and then
flips on his back, the way
a hand slaps hard on pavement.
I think of John: caught by power lines,
a flash exploding night, his dog
tags the only John we ever found.
I think of Susan: stabilizer separating, her
falling hopelessly through air.
I think of Mark: deadlocked eyes on target,
unable to unfreeze, turn or look away.
I think of you: flying up under the strain
of 5 g's, as tracers ignite and shells split
a calligraphy of anger. I think
of the night, three hours past

the Azores, Saint Elmo's fire
wrapped itself around the ship,
its green flame dancing on the windows
like small hard tears, the bolt
of green fire which shot
through the cockpit, the voice saying
'when you die it will be like this.'
Even now, on nights when I'm far
from the sound of any voice, I think
of you being there, saying *'No, this way.'* Then
it is your voice becoming mine. I am
a history of other voices, other hands at the controls;
a history of *'three green, no red or amber,'*
seat *'checked,'* pins *'pulled,'* canopy
'locked,' Caution Panel *'out:'* the day

the T-38 lifts from the runway, the pull
of afterburners, nozzles lit,
throttles which singe the air.
'No, this way' you tell the student
'Follow me through.' He feels
the tug of earth begging him back
as the plane slides up through seven thousand
feet as effortlessly as a raptor
arcs one wing before the pitch.
From the back of this tandem jet, all you see
is the dip of horizon, this slicing arc
of precision, the flash of sun
in your mirrors. All you feel is the blood
rush to your feet, the strain
of your pulsing heart against it.
'Fly the aircraft – or the aircraft flies you.'

The windscreen shatters at four hundred feet.
The pheasant whips through the front cockpit
and then splinters your canopy, hurled

at your face at three hundred knots.
The aircraft shakes:
could come apart at any moment
and you, left floating in clear air, nothing
holding; you, still holding
the stick, falling. In the street
below, kids are pointing at the sky.
Their mothers grab them and sprint
toward fields. Everything
is clear – *'your life or theirs:'* to know
such a pure and generous fear.
You tell the student to "punch."
He hesitates.
'Get out you son of a bitch!
Do it, do it now!' The blast of the front
rocket sled slams into you, blinding.
The plane turns south.
Beyond that, what else is there?
I speak to you, as I have
spoken every year
of my life. It has taken me
years to tell you this:
The last time I saw you, you were a tiny dot
disappearing over the neighbors' houses.
'There's still time' I hear you breathe
'Still time.'

for my father
for Tony Dater
for those I abandon

–P. H. Liotta

To Have Been

for Tony Dater, 14 January 1974

"Memory believes before knowing remembers."
–Faulkner

Mortality is just the difference
Between your blinking and your not blinking;
And if you blink at the wrong time in space,
Your eyes will freeze in a permanent blink,
In a frozen moment that will blind you
And bind you forever. But the last blink
Is the sum of all your blinks and becomes
You, you become history, history
Repeats, ... you are remembered. Memory
Believes and teaches us the past about
Ourselves: are we victims of our own
Proud knowledge? Or could the same hand create
Both the tiger and the lamb? Or bring
Together a white heal-all, white spider,
And a white moth? Or converge pheasant and plane
With nowhere to go except down, down, down?
Determined seven million years before
The sea did roll, was this day when the seat
Was saddled to your back. A take-off roll,
Gear up, bird on the wing – your own hand set
In motion Providence. You, too, had blinked
At it all before, but today it was
Different. You knew what we remember:
To have been is worth the being, Tony.

–James A. Grimshaw, Jr., Captain USAF

From a Pilot

for Walton F. Dater, LtCol, USAF (1932–1974)

Once, as I remember, we stood two miles above the sea
in a high mountain meadow, watching a hawk–
an eagle, maybe– wheel and soar. You said, "I see...
"Christ ... a windhover." No need for further talk.

So many such feelings are shared by those who fly:

Runway's end approaching fast on summer days;
slight sink from takeoff flaps just raised;
airborne safe sense when the airspeed builds;
thumps of turbulence; serenity of night time silence;
heart's clutch at red warning lights;
radio chatter; static crackle; leg straps' knife pain;
helmet's burn; misery of a too tight mask;
the welcoming beckon of a blinking strobe
through gray mist; then the roundout ground swell
and the abrupt, so satisfying feeling of the earth again.

Do birds sense this too? Or was Auden* right?
Perhaps they do live out their birdy lives
unconsciously indifferent to the fact of flight,
as are so many now, even some who fly. I've
wondered about this– and so, I know, did you.
Yet I wonder too what those pheasants felt to see
your strangely stiff-winged white jet hurtle through
their gabbling flock. They, like you and me
when flying, were no doubt going somewhere too.
But you and some of them fell yesterday
in a welter of feathers and fire

as once did Icarus, another man who knew
the wonder–and the sad, cruel beauty–of the sky.

–John Clark Pratt, 1974

*W. H. Auden, "Musee des Beaux Arts." Icarus was the name of the USAF Academy English Department literary publication formed by Colonel Dater.

To a Pilot's Widow

I know you so well: you were the wife I am today
– yesterday.
Today your life is gravely wounded, and, my sister
mine is too.
I dare not say, "I know how you feel,"
But just yesterday our concerns and hopes were
the same.

We married men who lived to fly and took such pride
in silver wings;
Our lives were bounded by their pride and sometimes
by their sorrow.
And we laughed with them and their little boy eyes
who thought their wings made them men apart –
And we knew they were.

We have counted the days together, you and I, existing
through each day, each hour until they were home.
We have tried to comfort some other one among us
Whose pilot came back under flag.

In the unreal today it is you, my friend, my sister,
holding the folded flag
The grief in your eyes stabs me and I have no words
to soothe so much sorrow.

And the todays yet to come, as my own pilot leaves secure ground for vast sky,
I will think of you and hope you have learned to live again,
And, for your brave, young husband, I will silently light a valiant candle
To burn there ... among my other prayers ... in the inner chapel
Where we have learned to lay down our fears and pray.

–Eileen Lundin

For Dawes, on Takeoff

I could have died
winding down the caprock road at night
driving hard to reach my father's bedside.
Canyon shoulders like the best vows
cave in, caliche ruts that crunch
soft as quail bones.

I drive black roads for bread, now,
tar roads soft all summer.
They whine on tires that spin me asleep.
I might die asleep, the calm habit we have
of saying of uncles, he slipped away quietly.

I could have died like Dawes,
suddenly inverted
two hundred feet on takeoff,
ailerons hooked up so both went up,
or down. He must have taken off
by faith. And rolled,
in spite of everything he tried,
inverted. Head down, yawing,

jerking the stick,
he must have squeezed the last controls
still working, ejecting downward like a dart,
and why not, about to crash anyway.

They found his head
stuck to his helmet.
In trees downrange, they found his arms
flung out from the body as if asking why.

–Walt McDonald

The Choice

I chose the skies
That few have known
To follow where
The winds have blown

To battle storms
That none have seen
To find seas of gold
And lands still green

And if some day
I don't return
Don't cry for me
Don't be concerned

For high above
The clouds will sing
For a world I loved
And silent wings

–Geoffrey H. Tyler

The Lady Let Him Fly

Never once
did she bind his wings;
take away his boyhood
paper-airplane-dreams;
nor try to force him
down to earth
when it was the air and sky
that beckoned his worth.
Never once
did the lady
hold him back,
or trounce his joy
for an air-to-ground-attack;
nor weep like a spoiled child
when he ventured into the blue wild.
In the background she would wait
chasing away twinges
for her fighter pilot's fate.
With wings straight and unfurled
he and the titanium bird
lifted above the runway's end
seeking freedom on the wind.
And when he did not return
the lady waited proud and strong
knowing he'd been – "happy all along."

And when the aged hands of Father Time
called him home
beyond the sky,
the young flyer smiled
because the Lady Let Him Fly.

–Kathleen M. Rodgers

Celestial Flight

She is not dead –
But only flying higher,
Higher than she's flown before,
And earthly limitations
Will hinder her no more.

There is no service ceiling,
Or any fuel range,
And there is no anoxia,
Or need for engine change.
Thank God that now her flight can be
To heights her eyes had scanned,
Where she can race with comets,
And buzz the rainbow's span.

For she is universal
Like courage, love and hope.
And all free, sweet emotions
Of vast and godly scope.

And understand a pilot's Fate
Is not the thing she fears,
But rather sadness left behind,
Your heartbreak and your tears.

So all you loved ones, dry your eyes,
Yes, it is wrong that you should grieve,
For she would love your courage more,
And she would want you to believe
She is not dead.
You should have known
That she is only flying higher,
Higher than she's ever flown.

–Elizabeth MacKethan Magid

Flying West

I hope there's a place, way up in the sky,
Where pilots can go, when they have to die.
A place where a guy could buy a cold beer
For a friend and a comrade whose memory is dear.
A place where no doctor or lawyer could tread,
Nor a management-type would e'er be caught dead!
Just a quaint little place, kind of dark, full of smoke,
Where they like to sing loud, and love a good joke!
The kind of a place where a lady could go,
And feel safe and secure by the men she would know.

There MUST be a place where old pilots go, when
Their wings become weary, when their airspeed gets low;
Where the whiskey is old, and the women are young,
And songs about flying and dying are sung.
Where you'd see all the fellows who'd "flown west" before,
And they'd call out your name, as you came thru the door,
Who would buy you a drink, if your thirst should be bad,
And relate to the others, "He was quite a good lad!"

And then thru the mist you'd spot an old guy
You had not seen for years, though he'd taught YOU to fly,
He'd nod his old head, and grin ear to ear,
And say, "Welcome, my son, I'm pleased that you're here!
For this is the place where true flyers come,
When the battles are over, and the wars have been won;
We've come here at last, to be safe and afar,
From the government clerk, and the management czar,
Politicians and lawyers, the Feds and the noise,
Where all Hours are Happy, and these good ol' boys,
Can relax with a 'cool one', and a well deserved rest.."
"This is Heaven, my son: You've passed your last check!"

–Michael J. Larkin

Last Flight

I went to see Bill last night
Lying in his flag-draped coffin
Flight plan in his pocket
Blue cap in hand
And a smile on his face
An Eagle Scout
He'd been prepared
To live in jungle's tangled growth
His unit decimated
Cut off from friendly lines
Two weeks alone
Among the enemy
He survived
He gave back to others
What had brought him through
Guiding tenderfeet
On Scouting paths
My son knew his tough love
But it was the love of flight
That drew Bill in the end
The dream to build
As first flyers built
A fragile craft
An ultra-light
To take him into the sky
I drove below that afternoon
Not knowing it was he
I watched the plane
On his last flight
Chugging a low and ragged arc
Above suburban sprawl
Was it a gust of wind
That drew him down
And slammed him into earth

Was he prepared
Did he know
The outer limits of his craft or man?
He was a first flyer
Daring the unknown

–Judy Humphrey, November 1996

The Path We Choose

Do not shed a tear for me
For I would not for you
Instead just drink a beer for me
And know well that I knew
Dreams of flight do not come free
There comes attached a price
And we do not do it blindly
We know we roll the dice
Before you sail into the sky
A sky slow to forgive
Answer am I, afraid to die?
Or just afraid to live?
So if you try, to reason why
When fate can seem unjust
We take these risks not to escape life
But to stop life escaping us.

–Dr. James Freeman

Death of a Flyer

We last saw him crouched,
Looking more like a giant animal in green fatigues,
A wobbling, drunken bottle,
Than a young, blond flier
Feeling his own blood begin to raise in mid air.
We laughed when he swung his lithe wife wildly
In the close August heat,
Rocking our garage ballroom,
As if the death whisper
Deep in his frail chest
Was not what oddly cocked and opened his face
Like a storm-battered door.
No one ever saw Billy Young dive in a sky-deep hole,
Floating above earth's small beings,
Leaping from man-skin
Into the breath-like moves of steel-skin,
Death blazing a ring of fire
Like sunlight from his fingertips,
Searing all life beneath him.
Billy Young, soft as candle wax
Burned in a blue eye of honor beyond all vision,
Leaving us forever frozen in a dancing, drunken room,
Our earth-bound bones swaying,
Too old, too tired, too slow
To rise from this grief's grave.

–David J. Smith

Untitled

Someday we will know, where the pilots go
When their work on earth is through.
Where the air is clean, and the engines gleam,
And the skies are always blue.
They have flown alone, with the engine's moan,
As they sweat the great beyond,
And they take delight, at the awesome sight
of the world spread far and yon.
Yet not alone, for above the moan,
when the earth is out of sight,
As they make their stand, He takes their hand,
and guides them through the night.
How near to God are these men of sod,
Who step near death's last door?
Oh, these men are real, not made of steel,
But He knows who goes before.
And how they live, and love and are beloved,
But their love is most for air.
And with death about, they will still fly out,
And leave their troubles there.
He knows these things, of men with wings,
And He knows they are surely true.
And He will give a hand, to such a man
'Cause He's a pilot too.

–Author Unknown

Nick

I still recall the night last year
The mate upon the phone
Something in the voice I hear
Just chills me to the bone
You hear 'bout Nick Dillane today?
Comes crackling down the line
Such simple words, said in a way
I knew he wasn't fine
It seems that he was towing
In winds a little strong
And there's no way of knowing
Exactly what went wrong
I hear the words wash over me
And a tear grows in my eye
Lockout...downwind...no release
No chance to say goodbye
Well it's nearly been a year today
Since I've seen his smiling face
Still seems like only yesterday
That he stood here in this place
Well I know that I won't see him
And I know that he is gone
But as long as I'm still breathin'
His memory will live on
So now its time to charge your glass
And just be glad we met him
Let's drink to Nick, whose time has passed
I know I won't forget him.

–Dr. James Freeman

Untitled

Ah design. That pheasants in their
twelve pound feathered glory
can bring to earth so many tons
of silver turbined thrusting might
so quick so sudden ... to eyewitness
unbelieving awe so numbing final.

A moth colliding with a toddler's knee
could not bring to earth so small a quarry.
How then a gnat a giant?

Design. That a hurtling thunder whine
sky devouring exploding furnace
should choke on pulp and feathers,
its windscreens blackened by ejection blast
and impact boiling bird blood,
the young saved and master dead.

A flock of pheasants where they never were,
intercepting death in an eyeblink flash.
Who can time such things?

Design. That a never thought of moment
leaprushed upon this instant of altitude,
a finger flick on the controls,
thirty feet, no more, of sky
rich with blackblur dots of bird ...
oh, what a migration, Lord, was that.

Trajectory closes with the whirl of horizon,
student chuting free to life below.
But what laid the farmhouse ahead?

Design. That crippled Icarus,
gravity drawn on leadrock wings,
still should bend his chattered craft
till nothing lay ahead but field,
plummet bailout, and, going gentle
even then, a windrush smile.

Bitter, Lord? No, furious.
Your design I, grieving, charge . . .
if design govern in a thing so large.

–Frederick T. Kiley, Lt Col USAF

In memory of Tony Dater, Lt Col USAF, 1932–1974.

Untitled

You tell me he died
instantly.
I do not believe you.
That would be too easy.
He would not die easily.
He would not know how to,
For he would have never done it before
in his life.

–Whitney I. Blair

Stand By

Listen Quietly

Be not disturbed by noises overhead
Made by a jet or plane of other kind;
Hear them with thankfulness to God instead,
Listen with calmness and a quiet mind.
Whether through boiling cloud or tranquil blue,
Throughout the dark of night and light of day
An Air Force pilot watches over you
Guarding your home, your life, your right to pray.
So, with a grateful heart and peaceful mind
Feeling secure beneath the plane above,
Pray for the pilot's safety, and you'll find
The noise has gone, and in its place there's love.

–Everett Milstead

Buzzing in a Biplane

Rocking the wings like drunks in a rowboat,
we give them something to think about
down there, dodging,
grabbing their heads on the midway,

the roar of our radial engine
drowning the lies of the barkers,
freaks at the sideshows
halting their acts and gawking.

Pulling up, we climb for the dark
like a roller coaster
loose from its tracks,
twisting a tear-drop turn

back at them, our blades
flashing fast in the night sky,
aiming at girls on the Ferris wheel
watching us dive

and screaming, wheel stalled
above the wings of our Waco.
Banked between girls
and high wires, we hope

all the gang is down there
choking on hot dogs,
holding their girls' hands
wet and trembling. We give them thrills

no carnival ever dared,
no sideshow this shocking,
two of their own in an open cockpit.
They won't forget us,

all of them screaming and pointing,
streaming like sheep in a stock pen,
the pony ride stampeding, ticket lines
swaying intact like a conga.

If we had the power
we would buzz them all night,
climb toward the sun blazing
so bright we'd blind them.

–Walt McDonald

Westbound Seven-Four

Here we sit
Galloping near silently
In the thin band between earth and space–
Toward a sun stranded
In the western sky.
Behind us,
Three hundred tons
Of metal, plastic, wire,
A small sea of fuel,
A brand-new Benz
And 320 souls,
Each with lives and purposes
Of their own.
And continents slide slowly,
Invisibly
Beneath our wings,
And oceans heave,
Their cold, black hearts unforgiving,
And we race, untouchable,
Just Mach shy,
Before the frigid wind.
Insulated in our skyborne city,
Flung like a dart
By some unseen giant.
Skipping on the exosphere
Where time dissolves
And dreams are made.

–*James MacNutt*

Dead Reckoning

At twenty-five thousand feet we are the distance.
Here there are no seasons for the blind: the only road
maps to the cities are the cracked blue veins
in ice that splinter off toward each horizon, the only life
in the lights of the oil rigs of Prudhoe Bay
five hours behind us. At night we fly through the aurora.
Green ghost arms weave themselves around the ship.
A sense of what seems right pulls us down to a pool of
phosphorescent shadow; the pilot's gyros betray us
straight, level. Here the magnetic compass turns
aside. The sun rises by falling south.
We imagine what we must: dead reckoning.
To fly by the stars, you follow what each
instrument predicts, even the idea
that the world flies out of an artificial horizon.

–P. H. Liotta

Reunion

I opened a box the other day
that I'd found on a closet floor.
Inside were the helmet and goggles
that I once so proudly wore.

As I pulled the helmet onto my head
it turned into a key
To the door of an airplane hangar
on the field on my memory.

I saw inside this hangar
the most beautiful things I've known.

Lined up was every airplane
that I have ever flown.

There was the Jenny and Swallow and Eaglerock,
and Waco and little Jay Three;
The Stitson and Beechcraft and Howard
were sitting there waiting for me.

As I petted old Jenny she sang to me
once again with the wind in her wires;
More seductive than any lover I've known
who'd excite my secret desires.

She said, "Come with me! We'll fly once again,
and I'll give you back your youth.
We'll circle and soar, and be joined once more
in a mating of absolute truth."

"We'll see the sights that eagles see;
feel the warm air's cushioned lift.
To travel the sky like the wild birds fly
is surely God's greatest gift."

"I'll anoint your face with warm oil
spread by my rocker arms,
And whisper to you while we're gliding
to land on some waiting farm."

But I said to her, "Dear Jenny
though I ache to go out and fly,
I have promised another lover
to be faithful until I die."

"This other love has no feeling for flight,
so I'll stay with her here on the ground.
She's given me such a share of her life
that to honor my promise I'm bound."

"So wait just a wee bit longer
in your hangar of memory,
'Til the day that I come and we'll fly once again
through the skies of eternity"

–Jim Leicher

Memory Patrol

My skies are always filled with ships
... Vespers of broken wings ...
Wheeling through cadences of time
In deep concentric rings;
Their landing fields the hangared years
From which they rise to fly
Each time I lift my wistful face
To contemplate the sky.
They fuel at memory's brimming tank
Without a pause in flight,
And drift with motors throttled down
Across my dreams at night;
Their cocards and their crosses blend,
For time has taken toll
And there is neither foe nor friend
In memory's patrol.

–Gill Robb Wilson

Biographies

Jeffrey C. Alfier is a Major in the United States Air Force. His poems have been published in Italian, Portuguese, and German. He earned a Masters of Arts in Humanities from California State University and is widely published.

Doug Atkins is a lifelong aircrew member. He claims a few drinks may have triggered the sentiment for his poem!

Richard L. Barlow dedicates "The Seat" and "The First Time" to his father, Richard H. Larsh, who taught him to see the soul of a word!

Timothy S. Bastian is a pilot, aircraft mechanic, and flight instructor who flies aerobatics, seaplanes, and tail wheel aircraft.

Carolyn Berge's poem "Feeling Compassionate" has a meaning that's pure and simple … it's a love poem. A love for the sky, clouds, eagles, and for the pilot who showed Carolyn the wonder of it all.

John Ciardi (1916–1986) served as a B-29 gunner during the Second World War in the 73d Bomb Wing, assigned to Saipan. After the war he taught English at Harvard, and served in various capacities as an editor, translator, teacher, and critic. He is perhaps best known for his translation of *Dante's Inferno*. He would become the author of several volumes of poetry. There are over thirty aviation poems among his works, and included with this anthology are "First Snow on an Airfield," "P-51," "Return," "The Pilot in the Jungle," and "Visibility Zero." Many of these may be found in his outstanding compilation of WW II aviation poems, "Other Skies" (1947), which contains some of Ciardi's most lyrically intensive works such as "Saipan," "Death of a Bomber," "Elegy," and "Two Songs for a Gunner."

Pat Collins is the wife of astronaut Michael Collins, Command Module Pilot of the first lunar mission.

James Dickey (1923–1997) may have been best known for his novel *Deliverance,* from which a popular film was made, but he was, above all, a poet. An early collection of his work, *Buckdancer's Choice,* won the 1966 National Book Award for Poetry. In all, he published more than 20 books, including first novels and criticism. Dickey also wrote 16 poems about flight. His most famous aviation poems were: "The Fire Bombing," "Two Poems of the Air," "The Performance," "The Enclosure," "Jewel," and "The Liberator Explodes." These poems can be found in *The Whole Motion: Collected Poems, 1945–1992,* Wesleyan University Press, 1992.

James Dickey enlisted in the Army Air Corps in 1942 to serve in World War II. Though Dickey himself did not complete pilot training, he eventually became a P-61 radar operator. He flew 39 missions in the P-61 "Black Widow" with the 418th Night Fighter Squadron based in the South Pacific.

Dickey's poem "The Firebombing" was originally 1668 words arranged into 316 lines of poetry. Due to space limitations, selected verses are included in this anthology. Now, decades later, though the memory of the firebombing remains poignantly evocative, it yet bears an aura of unreality, as the speaker cannot ever imagine such terror at his own threshold. Amid his genteel life in post-war America there are no "ears crackling off / Like powdery leaves, / Nothing with children of ashes". Nevertheless, the dead speak.

In "The Liberator Explodes," we witness a B-24 bomber crashing upon landing, perhaps due to its Davis Wing. The kinship is deep, as Dickey poignantly reminds us, for these are not simply vanquished fellow airmen. Though their faces passed unseen by the watchers, in an instant they are forever bound as brothers "of parallel fire," purchased by a moment of terror. Research conducted by Major Jeffrey C. Alfier, USAF.

James Freeman is an emergency physician who flies light aircraft, helicopters, and hang gliders. He lives in Australia and represented his country as a member of the national hang gliding team at a number of international competitions and is ranked in the world's top 50 pilots.

Randall Jarrell (1914–1965) was arguably America's foremost poet of World War II and this poem is the most anthologized and possibly most famous aviation short poem about the war. Randall Jarrell served four years in the U.S. Army Air Force at airfields in Texas, Illinois, and Arizona. Having failed to qualify for ferry-pilot training, he enlisted as a private and became a link trainer and celestial navigation instructor. He did not serve overseas in a war zone. Jarrell wrote 14 poems on flight. His most popular flying poems are: "Losses," "Eighth Air Force," "Second Air Force," "The Machine-Gun," "The Dead Wingman," "Siegfried," "Burning the Letters," "Pilots, Man Your Planes," "A Pilot from the Carrier," and "The Learners." These poems can be found in *The Complete Poems* by Randall Jarrell, published in 1969 by Farrar, Straus and Giroux, LLC.

Clay Greager owns the The Last Flight Out, an aviation specialty store in Key West, Florida. Well worth a visit if in the area. Clay knows a lot of interesting local folklore and has a unique philosophy that is enjoyable listening to.

Jack Greene published more poems (21) on soaring than any other poet to date. Jack's wife Audrey contributed 11 poems to this Anthology in his memory. Jack died of cancer in 1981.

Gene Griener was a retired Navy Air Traffic Controller who wrote many poems on different subjects. This one came from his book, *The Knife Is Wood.*

James A. Grimshaw dedicates "March 10, 1966" to Major Bernard Fisher for his heroic rescue, which earned him the first Air Force Medal of Honor in Vietnam.

Patrick Hamilton wrote "Cloud Dreamers" at Texas Tech University, July 2000. Dedicated to the love of flight and the few who share his passion. This poem is an expression of Patrick's search for identity and a place to truly call home. He believes only the few who have flown know this feeling and that "home" comes from within yourself.

Hansel Herrera is an aircraft maintenance technician for a small cargo airline in Miami, Florida. He became an aircraft mechanic in 1997 and shares a passion for aviation that only a select few can relate to. He attained airframe and powerplant certificates at George T. Baker Aviation school in Miami. His flying experience is limited to three hours of flight training on a Cessna 150, but he loves to be an observer in any aircraft. As a technician, he'd rather leave the flying to the pilots and ensure that the aircraft are safe and reliable. He writes poetry as a hobby to help him cope with the stress.

Daniel Hibbard is 20 years old and a senior electrical engineering student at Virginia Polytechnic Institute and State University in Blacksburg, Virginia. He is an avid glider pilot holding a private pilot's license for gliders. "The Cloud" is his first published work, but he has been writing about our surroundings and nature since grade school.

Judy Humphrey, a resident of Manassas and Augusta County, Virginia, was born in Newport News, Virginia. She earned a Bachelor of Science degree in business education from James Madison University and is also an accredited land consultant. Currently employed by the Salvation Army for thrift store operations and as Christmas coordinator in Loudoun County, Virginia, she is membership chair/secretary of the Real Estate Aviation Chapter, National Association of Realtors. She is not a pilot but loves to fly. She won a poetry award from Charlottesville Weekly in Charlottesville, Virginia.

Captain Rick Kerti is a Northwest Airlines A-320 pilot with 14,000 hours of flight time. He has a Bachelor of Science degree from Florida International University in Miami, Florida. He's flown more than twenty different types of aircraft. He is an artist, poet, and lover of kids. The lure of natural beauty as seen from the air really got him hooked on flying. The infinite beauty of the changing pallet of the sky and the terrain as viewed from the lofty perch is something few people get to enjoy.

Captain Michael James Larkin, a retired TWA pilot who has written poetry, short stories, and one novel, is a graduate of USAF Pilot Training Class 61-F. After serving as a Combat Ready crewmember on the Strategic Air Command B-47, he joined TWA in July of 1964 and retired there as a Captain of the B-747. His poems and short stories have been published in the *Airline Pilot, TARPA, Poetic Voices of America, 1998,The Lancet,* and other periodicals. He is a single father of five grown children and lives near Kansas City, MO. "Flying West" was written in 1991 upon the death of TWA Captain Edward R. "Buddy" Boland. Buddy was a Senior Captain loved and revered by young First Officers. "Flying West" was first published in the February 1995 issue of *Airline Pilot* magazine.

T. P. Leary's "Wright Field, 1944: The Ready Room" evokes memories of the historical developments at Wright Field, Dayton, Ohio.

Lieutenant Colonel P. H. Liotta, USAF was a KC-135 pilot and professor of national security affairs at he Naval War College. He was also a Fullbright Fellow in the former Yugoslavia and attaché to the Hellenic Republic. He's a prize-winning and prolific author with seven books to his credit in a wide variety of areas, including foreign policy, international security, education, and poetry. His writings have been translated into seven different languages.

James MacNutt was born in 1951 in Tris Rivieres. Quebec, Canada of Scottish-Irish Norwegian ancestry. He grew up with airplanes in the garage and spent family Sundays watching, flying, and studying airplanes. He spent 28 years as an air traffic controller in Canada. He stopped flying in the mid-nineties due to a medical restriction. His passions are his beautiful Lizzie, his sons, aviation, writing, and the sky-sciences, in that order.

Elizabeth MacKethan Magid wrote "Celestial Flight" in memory of Marie Michell Robinson, who died piloting a B-25 at 19 years of age.

Elizabeth and Marie were classmates in the Women Airforce Service Pilots (WASP) class of 1943. Elizabeth penned this poem en route to Marie's Memorial Service and presented it to her mother.

John Gillespie Magee was born in Shanghai, China and spoke Chinese before English. His parents were missionaries, an American father and an English mother. He was educated at Rugby school in England and at Avon Old Farms School in Connecticut. He won a Scholarship to Yale, but instead joined the Royal Canadian Air Force in late 1940, trained in Canada, and was sent to Britain. He flew in a Spitfire squadron and was killed on a routine training mission on December 11, 1941. The sonnet above was sent to his parents written on the back of a letter, which said, "I am enclosing a verse I wrote the other day. It started at 30,000 feet, and was finished soon after I landed." Magee's parents lived in Washington, D.C., at the time of his death, and the sonnet came to the attention of the Librarian of Congress, Archibald MacLeish. He acclaimed Magee the first poet of the War, and included the poem in an exhibition of poems of "faith and freedom" at the Library of Congress in February 1942.

Walt McDonald served in the USAF as a pilot flying T-34, T-28, B-25, T-33, and T-39 aircraft. During his career he served briefly in Vietnam in 1969-70. He also taught English at the USAF Academy and received a Ph.D. from Iowa between tours at the Academy. He is the Texas State Poet Laureate for 2001 and Paul Whitfield Horn Professor of English and Poet in Residence at Texas Tech University. He's published eighteen collections of poems and one book of fiction with publishers including Harper & Row and such university presses as Massachusetts, North Texas, Notre Dame, Ohio State, and Pittsburgh. He has published more than 1,900 poems of which 120 are specifically about flight. Six of his poems are included in this anthology. Dr. McDonald wrote more aviation poems than any other poet researched by the editor.

Howard Nemerov (1920–1991) was one of the most prolific and respected American poets of the twentieth century. From 1942 to 1944 he served with the Royal Canadian Air Force, and afterward with the U.S. Army Air Forces as a bomber pilot until the end of the war. A graduate of Harvard, he later became a Poet Laureate of the United States, and earned the Pulitzer Prize in 1978. Throughout his work, he engaged his readers with the genius of an unwavering intelligence.

'The War in the Air' exemplifies the wit, irony, and wisdom that critics near unanimously credit Nemerov with.

Dave B. Nichols was born in Cape Breton Island, Nova Scotia. He has a B.A. in English literature and music. He is an actor, broadcaster, writer, sailor, equestrian, and skier. He currently resides in Toronto, Ontario, Canada.

Gary Osoba is President of the Foundation for the Advancement of Micrometeorological Soaring. At a young age, he demonstrated a fascination with flying things and spent many of his early years training falcons and other birds. He was an early pioneer in the hang gliding field, and then went on to fly balloons, fixed wing aircraft, sailplanes, ultra lights, helicopters, and turbine aircraft. His current flying activities center solely around various classes of sailplanes. His achievements in natural flight have been recognized and awarded by the Soaring Society of America, the National Aeronautic Association, the Smithsonian National Air and Space Museum, and the Federation Aeronautique Internationale. He presently holds more general category world records in gliders than anyone.

David Pedlow was born in 1945 and resides on a hillside looking east over the Shropshire Plain on the border of England and Wales. He had a varied educational and employment history that includes two degrees and a career bagging coal by hand in a Devon coal yard. He is currently a tax accountant. He flies a Powerchute Kestrel para-plane that can fly up to 10,000 feet and 29 miles per hour. When not flying or writing, he can be found doing voluntary work on the nature reserve outside his back door, or skiving off on holiday.

John Clark Pratt served 20 years as a pilot in the USAF flying over 3000 hours. He spent one year in Vietnam as a pilot and operations analyst in 1969-70, flying mainly T-28s with combat missions in nine different kinds of aircraft including T-28s and O-1s in Laos. He earned his doctorate in English literature from Princeton and taught at the Air Force Academy. He was instrumental in creating the Vietnam War Literature Collection at Morgan Library, a unique and world-renowned collection. In 1974 he retired from the USAF with an Air medal, Bronze Star, and Meritorious Service medal. Dr. Pratt currently teaches at Colorado State University.

Kathleen M. Rodgers is a freelance writer whose work has appeared in a variety of publications, including *Family Circle, Virtue, Ideals, Good News, The Albuquerque Journal, Fort Worth Star-Telegram, Air Force Times, Army Times,* and *Navy Times.*

Her poem "To Live to Fly" was written for Mike "Rhett" Butler, a life member of the Edgar Allan Poe Literary Society. Mike is a renowned pilot who personifies what being a fighter pilot is all about. "The Searcher" was written for Lt. Brad "Booger" Hachat, her husband Tom's "Science Project." "Last of a Breed ... 'An Old Hand'" was written for Dave "Pig Rat" Erickson, USAF/Ret., an eternal fighter pilot.

Patricia Rockwell is a flight instructor with airplane single engine land and seaplane ratings. She was awarded the Amelia Earhart Memorial Career Scholarship in 1980 by the 99's (International Organization of Licensed Women Pilots) to continue her education in flight instruction.

Tom Schollie was born, raised, and educated in Yorkton, Saskatchewan, where his rhymes were published in his high school yearbooks. He started flying lessons at age fifteen and soloed a J-3 on August 26, 1947. Tom attended the University of Saskatchewan and obtained his law degree in 1955 and practiced for ten years. He was appointed Provincial Court Judge in Saskatchewan, first at Yorkton and then at Saskatoon, where he completed his B.A. in 1972 and his B.A. Advanced in sociology in 1982. He left in 1978 to serve as executive director of the Legal Aid Society of Alberta; in 1980 he was appointed to the Provincial Court of Alberta, where he still serves. In 1980 he took up gliding with the Edmonton Soaring Club and holds a Silver C badge. As a tow pilot, he has done over 2,200 tows, has 1,800 hours power, and over 300 hours in gliders.

Victoria Schrauwen wrote "Free" shortly after receiving her glider wing through the Royal Canadian Air Cadet Flying Scholarship Program. The poem represents her feelings about the magic of soaring.

Betty M. Simpson is a marketing representative, private pilot, and mother of three daughters. She published her first poem in 1979.

Ford H. Smart's "What Is a Fighter Pilot?" was found on a plaque hung on the wall at the 3247th Test Squadron Stag Bar, Eglin AFB, FL. This may not qualify as poetry, but the editor believes it is the best description of the fighter pilot mystique he has come across in many years of research.

Sir Stephen Spender (1909–1995) was an English poet, critic, lecturer, and professor. His writings often reflect democratic leftist and idealistic qualities, and he is perhaps best known for his contribution to *The God*

That Failed (1949), a volume of essays by those disillusioned with communism. Though he often romanticized pre-war Germany through his novels and translations of Hölderlin and Schiller, he expressed a fierce hatred of National Socialism. During the Second World War he served in the London Fire Service, and many of his poems reflect his experience amid the flames and destruction that resulted from the German blitz against London. *To Poets and Airmen* is a prime example of the poet's autobiographical reflections, being a powerful elegy on the iniquity of bombing innocent civilians and the dangers faced by a city's defenders and rescuers.

Eric Stice is a retired Air Force Reserve Lieutenant Colonel. (U.S. Air Force Academy Class 1970), a private pilot since 1965, and sometimes flight instructor in Alaska. Flying is his favorite hobby.

Karl Stice is a piano tuner by trade who lives in France. He's been passionate about ultralight aviation (hang gliding, paragliding, and trike flying) since 1974.

John Townsend Trowbridge (1827–1916) wrote the oldest poem found during my research. It was published in 1869, 34 years before the Wright brothers flew at Kitty Hawk. The original form contained 6 pages that the editor condensed to this version.

Raymond B. Tucker, Lt Col USAF's "The Fighter Pilot – a Tribute" is adapted with modifications from the *Aggressors Song Book*. This is not poetry, but it is among the best inspirational prose on fighter pilots found by the editor.

Geoffrey H. Tyler made an emergency landing in Angola among communist troops in 1981, while ferrying a small airplane from Florida to South Africa. Six weeks of intensive interrogation and two years of imprisonment followed from 1981 to 1983. This poem was written while in solitary confinement during his 22 months as a prisoner of the Communists in Angola. The poem was smuggled out by a Portuguese and forwarded to his mother, who had it copyrighted in 1982. Before imprisonment, Tyler flew for the U.S. Army and earned a Distinguished Flying Cross.

Patty Wagstaff is one of the world's most accomplished aerobatic pilots. She was the U.S. Unlimited National Aerobatic Champion in 1991, 1992, and 1993. This poem is from her book *Fire and Ice, A Life on the Edge* written by Patty Wagstaff with Ann L. Cooper, published by Chicago Review Press. One of the best books I've read and a truly interesting personality!

Gill Robb Wilson (1893–1966) was a legendary New Jersey aviation advocate who promoted aviation with the enthusiasm of an evangelical preacher. An analogy rendered even more fitting since he was also an ordained minister! He conceived the Civil Air Patrol in the late 1930s by foreseeing aviation's role in war and general aviation's potential to supplement America's unprepared military. Wilson, an aviation editor of *The New York Herald Tribune* and later New Jersey Aeronautics Commissioner, played a major role in the creation of the Civil Air Patrol. Gill Robb Wilson published over 50 aviation poems in several books about flight. He created the first aviation poetry book in 1938 called, *Leaves From An Old Log,* published by American Aviation Associates, Washington, D.C., 1938. His most famous book, *The Airman's World,* was published by Random House, New York, in 1957. His poem "Test Pilot" was written for the Society of Experimental Test Pilots.

Chris Woods is a film director who has directed more than 350 television commercials working through Atlas Pictures in Santa Monica, California. He is directing and producing the IMAX film, *Airspeed*, a drama based on the Reno Air Races. He was second unit director on *Forrest Gump, Michael, Private Parts,* and the soon-to-be-released, *The Sum of All Fears,* based on Tom Clancy's bestseller. He also wrote, produced, and directed the soaring films *The Quiet Challenge* and *Running on Empty* and has won numerous awards for his film work. He loves to write, since it's a big part of filmmaking. Flying has been his passion ever since he lifted off in a glider 25 years ago. He's logged over 3,400 hours (800 in sailplanes) and holds a commercial/instrument rating. He's a category-one competition soaring pilot and owns a Cessna 185, a restored 1933 Waco UBF-2 biplane, and a Schleicher ASW-27 sailplane.

William Butler Yeats (1865–1939) is considered one of the greatest English-language poets of the twentieth century. He wrote "An Irish Airman foresees his Death" to honor the death in the First World War of Major Robert Gregory, son of a rich playwright. There are strong undertones of Irish nationalism in the poem, and it is difficult to interpret it without this consideration. To what extent Yeats vicariously projects his beliefs onto those of Gregory is unknown.

Other Sources

of Aviation Poetry

Aviation Poetry Web Sites

- http://www.midwintercanada.com is the largest and most diverse site with 130 poems by 66 poets in four languages. Stewart Midwinter an internationally renowned hang glider pilot launched this site.
- http://www.landings.com/_landings/pages/poetry.html is the largest aviation Web Site containing 107 poems under the Hangar Talk directory launched by Ken Jenks of Mind's Eye Fiction who helped develop NASA's first aviation server.
- http://www.dynamicflight.com.au/Reading?Articles/poems.htm is the Australian Hang Gliding Web Site containing 14 poems.
- http://www.airforcehistory.hq.af.mil/soi/milavpoet.html is the United States Air Force index of military aviation poetry containing 5 poems and their history.

Aviation Anthologies

- *The Poetry of Flight, An Anthology*. Edited by Stella Wolfe Murray, 144 pages, published by Cranton Heath Limited, London, 1925.
- *Winged Ships*. Compiled and published by Eleanor Dixon Booth, Boston Massachusetts, 1925.
- *The Spirit of St. Louis: One Hundred Poems*. Edited by Charles Vale, 256 pages, published by George H. Doran Co, 1927.
- *Icarus: An Anthology of the Poetry of Flight,* by Rupert de la Bere, 191 pages, published by Macmillan & Co, London, 1938.
- *The Poetry of Flight: An Anthology*. Edited by Selden Rodman, 190 pages, Granger Index Reprint Series, published by Books for Libraries Press, NY, 1941.

- Poems From "Weekly Briefing." American Red Cross Officers' Club, Foggia, Italy, May 1945.
- *Puptent Poets of the Stars and Stripes Mediterranean.* Compiled by CPL Charles A. Hogan and CPL John Welsh, III, edited by LT Ed Hill, Italy 1945.
- *Listen. The War. A Collection of Poetry about the Viet-Nam War.* Edited by Lt Col Fred Kiley and Lt Col Tony Dater, 130 pages, published by the Air University Press, 1973.
- *Moonstruck: An Anthology of Lunar Poetry,* by Robert Phillips, Vanguard Press Inc, 1974.
- *Full Flight: An Anthology of Wooster Poetry Society of Ohio,* published by Wooster Poetry Society, 1977.
- *Skywriting, An Aviation Anthology,* by James Gilbert, 120 pages, St. Martin's Press, 1978.
- *Kites Flying: An Anthology of Poetry,* 44 pages, published by the Highgate Society, London.
- *First Flight: Poems (1960–1984),* by David Hunter, 80 pages, published by Gold Star Press, 1985.
- *The Flying Change.* Poems by Henry Taylor, 55 pages, published by Louisiana State University Press.
- *Thoughts Take Flight: An Anthology of Poetry and Stories about Airplanes, Pilots, and Flying,* by Allan and Barbara Stanely, 94 pages, CAVU Press, Harrison, NY, 1986.
- *Let's Pretend: Poems of Flight & Fancy.* Edited by Natalie S. Bober, 64 pages, published by Viking Press, 1986.

Aviation Poetry Books or Sources

- *Leaves From An Old Log,* by Gill Robb Wilson, published by American Aviation Associates, Washington, D.C., 1938.
- *Wind, Sand and Stars,* by Antoine de Saint-Exupery, published by Harcourt Brace Jovanovich, NY, 1939.
- *The Airman: A Poem in Four Parts,* by Selden Rodman, 148 pages, published by Random House, NY, 1941.
- *The New Treasury of War Poetry: Poems of the Second World War.* Edited by George H. Clarke, published by the Riverside Press, Cambridge, MA, 1943.
- *Other Skies,* by John Ciardi, published by Little, Brown, Boston, 1947.
- *Sky Argosies in Seven Missions,* by De Jean and Louis Leon, 25 pages, published by Sky Argosies (Orange, NJ), 1951.
- *The Airman's World,* by Gill Robb Wilson, published by Random House, NY, 1957.
- *Favorite Poems, Old and New, Selected for Boys and Girls,* by Helen Ferris, published by Doubleday & Co, 1957.
- *Buckdancer's Choice,* by James Dickey, published by Wesleyan University Press, Middletown CN, 1965.
- *The Complete Poems,* by Randall Jarrell, published by Farrar, Straus & Giroux, NY 1969.
- *Man in the Sky: Modern Poetry of Flight,* by Malcolm Ford, 42 pages, Vantage Press, 1970.
- *Carrying the Fire,* by Michael Collins, published by Farrar, Straus and Giroux, 1974.
- *Icarus: A Magazine of Creative Writing,* published by Dept of English, USAF Academy, 1974.
- *The Strength of Fields,* by James Dickey, published by Doubleday & Co, NY 1979.

- *Anthology of Magazine Verse and Yearbook of American Poetry, 1980 Edition,* by Alan Pater, published by the Monitor Book Co, 1980.
- *The Hopwood Anthology, Five Decades of American Poetry.* Edited by Harry Thomas and Steven Lavine, published by The University of Michigan Press, 1981.
- *The Wild Blue Yonder: Songs of the Air Force.* Edited by C. W. "Bill" Getz, published by The Redwood Press, 1981.
- *The Random House Book of Poetry for Children.* Selected by Jack Prelutsky, published by Random House, NY, 1983.
- *Soul Flight,* by Emily Powell, published by Pittenbruach Press, 1986.
- *The Wild Blue Yonder: Songs of the Air Force, Volume II, Stag Bar Edition.* Edited by C. W. "Bill" Getz, published by The Redwood Press, 1986.
- *The Flying Dutchman,* by Walt McDonald, Ohio State University Press, 1987.
- *Sunshine and Shadows of Soaring,* by Jack Greene, 21 pages, published by Aero Club Albatross, 1989.
- *The Eagle's Mile,* by James Dickey, published by Wesleyan University Press, Middletown CN, 1990.
- *Test Flying at Old Wright Field.* Collected by Ken Chilstrom, Edited by Penn Leary, and published by Westchester House, Omaha, NE, 1991.
- *Zeppelin Reader: Stories, Poems, and Songs from the Age of Airships.* Edited by Robert Hedin, published by University of Iowa Press, 314 pages, 1998, ISBN: 0877456291